Reconstructions of Gender and Information Technology

"An eye-opener regarding what it really takes to achieve gender parity in tech—not only in the Nordic countries but all over the western world. Based on extensive research, Hilde G. Corneliussen offers rich insights to understand why the assumption that gender equality is already in place can actually make things worse. And she delivers innovative ideas on what to do about it."
—Nicola Marsden, *Professor for social informatics, Heilbronn University, Germany*, author of *Retaining Women in Tech: Changing the Paradigm*

"Through firsthand accounts of Norwegian women's experiences in IT, Corneliussen reveals how gender bias can persist even in the world's most egalitarian societies. Refuting the assumption that interest in IT is an immutable gendered trait, her focus on women's motivations, choices, and sense of identity and belonging brings new insight into female participation in IT and how to increase it."
—Janet Abbate, *Professor of Science, Technology and Society at Virginia Tech, USA*, and author of *Recoding Gender*

"Hilde G. Corneliussen provides an exploration of gender equality in technology with this definitive and thought-provoking book. With meticulous research, she sheds light on the unsolved issues of the gender imbalance in IT, revealing the complex factors that hinder progress. Through compelling narratives and inspiring insights, the book unveils the resilience of women who challenge stereotypes and reconstruct the gendered space of IT. A must-read for those seeking to create an inclusive digital future."
—Jeria Quesenberry, Associate Dean of Faculty, *Carnegie Mellon University, USA*, and author of *Cracking the Digital Ceiling: Global Views of Women in Computing*

"Professor Corneliussen's highly readable book offers an innovative approach to a decades-long concern about the absence of women in IT. It is notable for her exploration of multiple and often unconventional pathways into IT, challenging the so-called Nordic paradox whereby gender inequality continues to prevail in this male dominated profession. This timely and important publication brings hopeful insights and practical advice for the future of the IT sector and for gender equality more widely."
—Clem Herman, *School of Computing and Communications, The Open University, UK*, and author of *Women in Tech: A Practical Guide to Increasing Gender Diversity and Inclusion*

Hilde G. Corneliussen

Reconstructions of Gender and Information Technology

Women Doing IT for Themselves

palgrave
macmillan

Hilde G. Corneliussen 🆔
Western Norway Research Institute
Sogndal, Norway

ISBN 978-981-99-5186-4   ISBN 978-981-99-5187-1   (eBook)
https://doi.org/10.1007/978-981-99-5187-1

Cover illustration: pattern © Melisa Hasan

This Palgrave Macmillan imprint is published by the registered company Springer Nature Singapore Pte Ltd.
The registered company address is: 152 Beach Road, #21-01/04 Gateway East, Singapore 189721, Singapore

Paper in this product is recyclable.

ACKNOWLEDGMENTS

The topic of women in technology has been close to my research heart for more than 25 years. The building blocks of insights that contribute to this book have developed through a series of research projects and in collaboration with researchers in Norway and internationally during this period, and I extend my heartfelt gratitude to all those who have been instrumental in making this book a reality. The three projects providing the data material for this book also involves collaborators, and I extend my sincere appreciation to my colleagues at Western Norway Research Institute, in particular Gilda Seddighi and Carol Azungi Dralega, who have been close collaborators and important discussion partners in many relevant projects and topics developed in this book. I also want to thank Anna Maria Urbaniak-Brekke and Morten Simonsen for collaboration on one of the projects, and I extend my thanks to Vibeke Valkner and Karolina Dmitrow-Devold for their assistance during the research process. I also want to thank all my colleagues involved in the Nordic Centre of Excellence on women in technology, *Nordwit*, led by Gabriele Griffin: thank you for the many engaging and interesting discussions.

I am also grateful to the funding organizations that have supported this work, in particular a heartfelt thanks to the National Centre for STEM recruitment for inviting to and funding the study that forms the foundation of the main narrative throughout the book as well as for the continuous dialogue. Thanks to Nordforsk the NCoE Nordwit could provide a solid research base over the last five years, and thank you to the Norwegian Directorate for Children, Youth and Family Affairs (Bufdir) for engaging myself and colleagues at the research institute for the evaluation of the Girls and Technology campaign.

I also want to thank the many informants and a wide range of contributors representing stakeholders from education, the public and private sectors, who have generously shared their experiences, knowledge, and viewpoints in the research process. You have made it possible to expand our knowledge by providing invaluable insights into how the message of gender equality and gender balance is received and understood, and how it can be transformed into actionable change.

Time and space are invaluable for the process of writing a book. Therefore, my heartfelt gratitude goes to Western Norway Research Institute for generously providing me with the isolated time needed to develop the different research projects into a coherent book. Thank you so much to Isabel and Tim Riley for providing space through their gracious invitation to my family, including my husband and our two always happy dogs to stay at Waveney Lodge for three months while I focused on writing the main part of this book. Their hospitality and support were truly remarkable. A special thank you also to Isabel for proofreading and to Tim for serving the best latte. My deepest appreciation goes to my husband for his unwavering support throughout this journey.

The six chapters of the book are written specifically for this book. However, since the material presented is developed through three research projects, data material and earlier analyses have previously been presented as follows:

Chapter 3 presents an updated and refined version of the material and analysis from the original project report in Norwegian:

Corneliussen, H. G. (2020), "Dette har jeg aldri gjort før, så dette er jeg sikkert skikkelig flink på"—Rapport om kvinner i IKT og IKT-sikkerhet, Sogndal: VF-rapport 8/2020.

Chapter 4 presents a comprehensively reworked version of a previous book chapter that has been reorganized and updated with new data material and a new theoretical framework for the current book:

Corneliussen, H. G. (2021). Women Empowering Themselves to Fit into ICT. In E. Lechman (Ed.), *Technology and Women's Empowerment* (pp. 46–62). London: Routledge.

The interviews with schools presented in Chap. 5 have been presented in a project report in Norwegian. The text in Chap. 5 includes additional data material and a new analysis within the theoretical framework of the book. My co-authors for the Norwegian report have given their consent to the re-use of the data material.

Corneliussen, H. G., Seddighi, G., Simonsen, M., & Urbaniak-Brekke, A. M. (2021). Evaluering av Jenter og teknologi: VF-rapport 3/2021.

I am grateful to Palgrave for their guidance and support in the process of making the manuscript into a book.

Thanks to the Western Norway Research Institute's funding, you can read this book as an Open Access title.

CONTENTS

ABOUT THE AUTHOR

Hilde G. Corneliussen is Research Professor in Technology and Society at Western Norway Research Institute. Her research and scientific publications are mainly on how to make technology more inclusive for groups at risk of being excluded from the digital transformation, including *Gender–Technology Relations: Exploring Stability and Change* (2011).

LIST OF FIGURES

CHAPTER 1

Women Fighting Gender Stereotypes in a Gender Egalitarian Culture

Abstract Despite the increased importance of technology, at current rates it will still take hundreds of years to achieve gender equality in technology across western countries, even in the Nordic countries, which are recognized as some of the most gender-egalitarian nations in the world. Challenging this situation requires knowledge about how women in fact come to participate in fields of information technology (IT), which is the topic of this book. This chapter presents the framework for the book, including the overview of the theoretical and methodological perspectives, the empirical material from a series of studies in Norway, and some of the relevant debates that the book engages in. The research presented here is rooted in a tradition of feminist technology studies and inspired by feminist studies of contexts where women face gender barriers and challenges in identifying their belonging.

Keywords Digital transformation • Gender disparity in information technology • Mystery of recruiting women to IT • Myth of gender equality • Nordic gender equality paradox • Space invaders

© The Author(s) 2024
H. G. Corneliussen, *Reconstructions of Gender and Information Technology*, https://doi.org/10.1007/978-981-99-5187-1_1

1

INTRODUCTION: CONTINUOUS GENDER DISPARITY IN TECHNOLOGY

Why is it important to study women's relationship to information technology (IT), when apparently *everybody* is into IT today? The traditional gender gap in the access and uptake of IT still found in many countries globally (Borgonovi et al., 2018) is not the most critical issue for the coming generation in Norway. 100% of young women aged 16 to 24 in Norway use technology to access the internet, and nearly all young people engage in digital media (Schiro, 2022; Statistics Norway, 2022). Yet there is no indication that this has improved the underrepresentation of young women choosing a career as IT experts. Digitalization and digital transformation across sectors, industries, and occupations make IT experts a highly valued resource. The demand for IT specialists grew nine times faster than the total increase in the European labour market between 2011 and 2020 (Eurostat, 2021b). The demand for IT professionals is also expected to grow significantly in the coming years (World Economic Forum, 2020a). IT specialists are, however, a scarce resource. Closing the gender gap in the IT sector and among IT experts is therefore not only necessary for gender equality and women's empowerment. It is also important for national economies that otherwise risk a substantial economic loss if they fail to bridge the digital skills gap within a few years (European Union, 2021; Palmer, 2021; Quirós et al., 2018).

While the importance of making the IT sector more inclusive by recruiting from a broader selection of people is widely recognized (Chavatzia, 2017), how to achieve such diversity seems to remain not only a mystery, but a challenge that many have given up on: "We have just been through a process of investigating what we can do to attract more women to our department. However, we concluded that there is nothing we can do because it is too late by the time the women apply to university." A professor in computer science at a European university told me this in a discussion about how to recruit women to IT degrees. His department wanted more female students; however, changing the current gender pattern among students applying for an IT degree appeared to be out of reach for them.

"Do we really need more women in IT?", one of the male managers in an IT company asked in a meeting about how to get more women into IT jobs (Corneliussen & Seddighi, 2020a). Although he and his colleagues wanted to welcome more women working as IT experts, few women applied for such positions. He assumed this had to do with most women lacking interest in IT, which seemed to justify a passive strategy and rather to accept that few women applied to IT jobs (ibid.).

"You have to remember that girls are not really interested in IT." The first time I took notice of this claim was in a group of educators who had been challenged to think about ways of getting more girls to join the programming class in 9th and 10th grade. After a full day of discussing how to approach the challenge, the headteacher still hesitated to take on the challenge of making the programming class equally popular among girls as it was among boys, doubting that reaching gender balance in the programming class would be possible. Girls' apparent lack of interest was not only considered the *main* problem, but was also regarded as *unchangeable*. I started to notice how the argument of girls' and women's lack of interest reappeared over and over, from educators, policymakers, employers, parents, IT professionals, men as well as women. Some of those who could and should have been first in line to support and motivate girls and young women to learn more about IT seemed to have lost all hope. The quotes above illustrate some of the core challenges for creating a more gender-inclusive and gender-balanced IT workforce: wrapped in good intentions about gender equality, a widespread distrust of girls' and women's interest in IT that might result in less efforts to recruit girls to IT.

In 2011 I referred to the "stability argument", such as the notion that "the underrepresentation of women has still not improved" (Corneliussen, 2011, p. 2), and I called for research on change and improvements, rather than focusing only on stagnation. Internationally, stagnation still seems to dominate the picture, depressingly underlined by the World Economic Forum's claim that it will take more than 250 years to reach gender parity in IT (Palmer, 2021; World Economic Forum, 2020b). It is time to explore new ways of speeding up the process of change, but also to recognize that the last decade has witnessed some improvements. There were, for instance, nearly twice as many women signing up for a university degree in information technology in Norway in 2022 compared to ten years earlier (The Norwegian Universities and Colleges Admission Service, 2022). But there are still large differences, and, as the following chapters will illustrate, there is still a long way to go before girls in general can approach fields of IT with the same naturalness as boys.

One worrying trend is the overwhelmingly large amount of research documenting continuous gender bias, negative attitudes and distrust of girls' and women's interest and skills in IT (Buse, 2018; Kenny & Donnelly,

2020; Yates & Plagnol, 2022). These attitudes make it difficult for women to engage in IT. This book will contribute to this field by examining how women navigate the gendered landscape of IT, not only by looking at barriers challenging their engagement, but also by exploring what makes women enter fields of IT. The context of the research presented here is Norway, one of the Nordic countries that for years have occupied the top positions of international gender equality rankings such as the World Economic Forum's *Global Gender Gap Index* (2020b). The score reflects the state as an active agent for introducing gender equality and gender mainstream in a wide set of spheres, from health and welfare, to wage policy, and a focus on representation and participation (Holst et al., 2019). The Nordic gender equality model is recognized as exceptional with its family-friendly working life policies that have provided an important foundation for women's high participation in paid work (Teigen & Skjeie, 2017).

Although nearly as many women as men are active in working life, there are still many examples of inequality such as more women working part time, women dominating public sector while men dominate in private sector (Statistics Norway, 2018). 60 % of students in higher education and nearly half the population of research personnel are women. There are, however, fewer women higher up in the academic hierarchy. An effort to improve the gender balance among full professors has resulted in an increase from 24 to 36% between 2012 and 2022. This is higher than the EU average (European Commission, 2021c), however, not in technology, which have not only the lowest percentage of women, but also, together with other science, technology, engineering, and mathematics (STEM) fields, the lowest levels of improvement over the course of the last decade (Olsen & Wendt, 2023). This reflects a more general picture of gender equality as a widely accepted norm and a goal, though with fields of technology lagging behind (Foss, 2020).

While gender equality undoubtedly represents an important value in the Nordic countries, as the numbers above suggest, there is still a notable gender inequality, marked by a vertical gender segregation with few women in top positions, and a horizontal gender segregation between educations and occupations. Although both these patterns appear to be in conflict with the ideal of gender equality (Ellingsæter, 2014; Sund, 2015), the image of the Nordic cultures as gender egalitarian remains strong, feeding a *myth* about gender equality already existing (Martinsson & Griffin, 2016). Gender equality thus appears here as a *description* of society, rather than a political goal. In international and global contexts, this works as a "nation branding", Larsen and colleagues suggest, and "as a political and symbolic value" that reinforces the Nordic countries "moral superpowers" (2021, p. 2). In the following chapters this will become evident in terms of how gender equality as a goal seems to go under the radar in contexts of IT, simultaneously as the highly valued gender-egalitarian norms remain intact, unscathed by the continuous underrepresentation of women in the field. This has often been labelled the Nordic gender equality paradox, accompanied by a finger pointing at women and their (poor) choices of education (Corneliussen, 2021a; Stoet & Geary, 2018). This is certainly a paradox; however, we need to look beyond women's educational choices to find the necessary answers and solutions to the challenge of gender equality becoming a *non-performative* policy, which is what Ahmed labels policies that "do not bring about the effects they name" (2012, p. 17).

What makes Women Enter IT?

For decades, it has been taken for granted that early tinkering and playing with computers have been important gateways in leading boys into professional IT careers (Sevin & Decamp, 2016). This has nourished a hypothesis of the low proportion of women in IT as a reflection of girls' and boys' different approaches to technology, assuming that boys acquire more hands-on experience than girls (Gerson et al., 2022). Initiatives to increase girls' and women's participation in IT have often attempted to mimic boys' interests, or simply teaching girls what it is assumed that boys already know (Corneliussen, 2011; Margolis & Fisher, 2002; McKinsey & Company and Pivotal Ventures, 2018; Sharma et al., 2021). The research that I will share in this book indicates that such initiatives have not reached all women and that not all schools work as active recruiting grounds for getting more girls to consider IT. An evaluation of national authorities' campaign to increase Norwegian youths' interest in science and mathematics also found that schools' efforts to recruit youth to science had not succeeded. Instead, the numbers had moved in the wrong direction (NIFU, 2021). Furthermore, what has been judged as successful strategies to recruit women to IT degrees at university have not proved to have long-lasting effects (Lagesen et al., 2021). A Norwegian Official Report documents that youth in Norway are still highly affected by gender norms when choosing education (NOU, 2019: 19). Our studies during the last decade have also shown that computing and IT are still coded masculine in Norway (Corneliussen & Prøitz, 2016; Corneliussen & Seddighi, 2020a), similar to the gender coding of these fields in most other western countries (Arnold et al., 2021; Barbieri et al., 2020; Holtzblatt & Marsden, 2022; Misa, 2010a). While many have asked why there are so few women in IT, it is equally important to ask how women who have pursued a career in IT have navigated the still active gender stereotypes of technology. This is the main topic of this book.

What has led women to IT, and what can we learn from them? These are among the questions guiding the forthcoming chapters. Here I will use the term IT to refer to various fields of IT education associated with faculties and universities of science and technology, and IT work or IT jobs referring to work that involves a similar type of competence.[1] The main narrative of the book follows women with a background in these fields. The term also captures a widespread pre-understanding among these women; before they enrolled in an IT degree at university, their lack of knowledge about different disciplines made their perceptions of this unspecified IT dominate their expectations, similar to patterns documented in other Nordic countries (Vainionpää et al., 2019).

The women who share their narratives here are engaged in fields of IT as students, early career researchers at universities, or workers in the IT sectors. Chapter 3 explores their experiences of navigating IT as a field new to most of them, from childhood to university. Chapter 4 analyses how the women approach IT mainly as fields dominated by stereotypes about men as well as women and of images of male IT experts, but also how they challenge and revise such gendered images. Women are not alone in navigating the landscape of IT. Chapter 5 changes perspective to consider how representatives for lower and upper secondary schools approach issues of gender disparity in IT and how they interpret their own role as a supportive arena for the young women. Interviews with school representatives provide insight into one of the critical arenas for recruiting young women to IT.

This book offers insights into how young women today outline new ways of understanding core values of IT as well as relevant qualities for working with IT. Some of the solutions tried and tested by the women (Chaps. 3 and 4) are different from what is anticipated by, for instance, schools (Chap. 5). Many young women depend on events and resources *outside* the traditional educational system, and many find alternative routes leading to IT degrees. The reconfiguration of IT that the women propose includes a highly diverse set of interests, competences, and skills suitable for a diverse set of people. Instead of trying to blend in and become invisible as women, they rather use their "token" identity (Kanter, [1977] 1993) as a tool for claiming visibility for women in IT. Thus, the women's narratives point towards a more gender-inclusive digital future by illustrating alternative pathways to IT and by reconfiguring images of gender and technology to fit a wide set of competences and interests leading to professional engagement within IT. It is time to explore how women are *doing IT for themselves* to learn more about how to develop successful strategies for increasing girls' and women's participation in IT.

Women's Underrepresentation—A Multidimensional Challenge

The challenge of the continuous underrepresentation of women in computing involves several overlapping, but not identical, issues. The *leaky pipeline* is one metaphor often used to illustrate the challenge of keeping women in the field once they have started. This metaphor has also been criticized for picturing a single, standard pathway to technology, which reduces our ability to capture how certain individuals find unconventional routes to IT (Corneliussen & Seddighi, 2022; Vitores & Gil-Juárez, 2016). Holtzblatt and Marsden, claiming that retention is the main problem, illustrate it with a "leaky bucket" that is constantly filled with women who then "leak out" of the IT sector to a higher degree than men do. This suggests that the IT sector largely fails to provide women with a work experience "as reliably as to men" (Holtzblatt & Marsden, 2022, p. 10). In this book, however, it is not the leakage, but rather the problem of filling the bucket, that will be explored, including the alternative pathways leading to IT. We will learn more about how girls and women experience that traditional recruitment channels such as schools and educational institutions, largely fail to provide them with what they need for considering IT as an equally natural choice as many young men do.

While we can see notable gender differences in this field, it is important not to exaggerate these (Hyde, 2005), as there are also large differences among boys and among girls—not all boys see IT as a "natural" choice, while some girls do. This book takes as a starting point that men and women experience some of the most gender-divided educational choices differently, reflected in a notable horizontal gender segregation within the Norwegian higher education sector (The Norwegian Universities and Colleges Admission Service, 2022). Here this is explored through, for instance, the women's narratives describing a rather uniform experience of

IT as a field populated mostly by men, and identified through gender stereotypes defining who belongs in IT. However, an equally important part
of the book is to explore the differences between the women: their experiences illustrate that they come to IT from different backgrounds, with
different types of support and motivation at various points in their chronological narratives. The variations between the women also suggest that
there is no singular solution to the challenge of increasing women's participation in IT. Their experiences can, however, work as guidelines for
ways of facilitating girls' and women's perception of IT as a relevant and
interesting education and career choice. Using Norway as an example provides the opportunity of exploring how a gender-egalitarian culture
(Teigen & Skjeie, 2017; World Economic Forum, 2020b) can combine
gender equality as a vital value in society with a continuous gender imbalance in education and work, particularly noticeable in fields of technology.
The knowledge developed in this book thus holds relevance to a wider set
of cultures that also experience a challenge of increasing women's participation in IT.

The rest of this chapter will present the theoretical and methodological
framework within gender and technology studies, then introduce the
empirical material and the analytical strategy, before laying out the structure of the book.

Theoretical and Methodological Framework for the Book

The basic understanding of gender guiding the analysis in this book is
gender as something we do rather than something we *are*. West and
Zimmerman's theory of "doing gender" acknowledges that gender identity is constructed in and through social relations and interactions in
social contexts (1987). Doing gender is an ongoing accomplishment, and
also one in which "members of society 'do difference' by creating distinc-
tions among themselves" (West & Zimmerman, 2009, p. 114). Although
such differences are not natural, once they have been established, they are
used to affirm and endorse practices that reproduce the same differences
(ibid.).

Technology is also affected by such differences and gender influences
criteria for being perceived as an IT expert. Across the western world, IT
has been perceived as a field "outside the female domain" (Borgonovi

et al., 2018; Trauth & Quesenberry, 2007). Researchers from feminist technology studies have for a long time emphasized that gender and technology are socially constructed and that they need to be understood in terms of how they shape or co-construct each other (Bray, 2007; Cockburn, 1992; Cockburn & Ormrod, 1993; Landström, 2007; Wajcman, 2010). Many studies have explored the challenging relationship between gender identity and IT for women, for instance identifying a narrative of IT as a "world without women" that makes women appear "out of place" (Sørensen, 2011, p. 45), and assumptions of "femininity and technical ability as virtually incompatible" that challenges even the most skilled women (Kenny & Donnelly, 2020, p. 343). Such perceptions have made women moderate their femininity (Wajcman, 2004), trying to blend in with a masculine culture (Turkle, 1988), and adopting a work style associated with men (Watts, 2009). Studies have found that women in technical roles have been perceived as different from other women (Kenny & Donnelly, 2020). Faulkner identifies this as a heterosexist discourse where "women who are really into engineering are not 'real women' and conversely 'real women' are not 'real engineers'" as a message that is constantly reproduced in engineering culture, practices, and identities (2014). Some women who are passionate about technology have embraced gender stereotypes, for instance claiming a similarity to boys to explain their "geekiness" (Bury, 2011). A larger body of research documents that many women find it hard to negotiate between the messages of "gender in/authenticity" identified by Faulkner (2014). Gender and technology are, however, not fixed in a static relationship. The relationship varies between women and has changed over time (Trauth & Connolly, 2021), though, not always in a positive direction. While women's contributions to computer programming and software developing in the post-war decades is a neglected field in modern narratives of computing (Abbate, 2012; Vitores & Gil-Juárez, 2016), recent research has found that women who have worked in IT give accounts of *fewer* women in such roles recently compared to when they first joined the field (Kenny & Donnelly, 2020). A recent report from one of the Nordic countries found, after years of monitoring how young girls perceive IT, a deteriorating trend in the girls' knowledge about what it means to work with IT (Insight Intelligence, 2022). The gendering of IT fields is not uniform, but differs across the world (Borgonovi et al., 2018; Castañeda-Navarrete et al., 2023), with a notably larger gender divide in the more gender-equal countries, while more authoritarian countries can identify a larger proportion of women in

IT education (Chow & Charles, 2019). This has led some researchers to suggest that this is a paradox resting on women's preferences for choosing gender-traditional educations and occupations (Stoet & Geary, 2018), a view that will be discussed and nuanced in this book.

Women have different backgrounds and interests that give them different motivations for choosing an education within IT (Dee, 2021). The Norwegian women's narratives document that their experiences reflect a culture in which gender stereotypes and cultural discourses still define women as outsiders in fields of IT. To understand women's encounters with the masculine cultures and gender stereotypes of IT, this book engages theoretical perspectives that capture the experiences of navigating discursive spaces while facing the challenge of fitting in.

Doing gender in IT often appears to be quite different for women and men, and the concept of doing gender has contributed to insightful studies of women negotiating gender identity in IT. It has been suggested that women find it challenging to combine "doing gender" with "doing IT" while facing cultural stereotypes that make women and technology appear as a "contradiction in terms" (Nentwich & Kelan, 2014, p. 128). Some women have experienced this as an "in/visibility paradox" (Faulkner, 2009), of being visible as women but invisible as professionals. West and Zimmerman's concept of the "if-can-test" points to how people use their cultural knowledge to categorize others. If you *can* identify someone with a specific category, you do so (West & Zimmerman, 1987). The "if-can-test" thus highlights how some identities are perceived to fit, while others do not comply with the basic features of a category. Masculine images of IT, for instance, are more challenging for women to negotiate than they are for men.

Nirmal Puwar's concept of space invaders elaborates on how certain bodies do not feel welcome in certain spaces (2004). The analysis of how women navigate the gendered spaces of IT is inspired by Puwar's study of women and racial minorities entering spaces that are rarely occupied by them (2004, p. 141). Puwar shows how previously excluded groups entering spaces that have been historically or conceptually not "reserved for them" capture a moment of change and disturbance of status quo. She develops this into a technique for exploring "how spaces have been formed through what has been constructed out" (Puwar, 2004, p. 1). The concept of "space invader" is borrowed from Doreen Massey (1996), who used this concept to explain her experience of being a woman in male-dominated spaces of football. The paradox in Massey's reflection, of being there, enjoying it, but not fully belonging (Massey, 1996, p. 185), is at the core of Puwar's study of women and racial minorities entering historically male and white spaces (2004, p. 2). Space invaders are "bodies out of place", and the challenge highlighted by Puwar is the visual signs of not belonging within the bodily or "somatic norm",[2] highlighting the frictions of "the increasing proximity of the hitherto outside with the inside proper" (ibid., p. 1). The space invader thus captures the double position that, for instance, women in male spaces of IT can have; by not representing the bodily norm "they don't have an undisputed right to occupy this space. Yet they are still insiders" (Puwar, 2004, p. 8). Male bodies have often been taken to represent "empty, neutral positions that can be filled by any(body)" (Puwar, 2004, p. 32). Bodies that fit the norm can pass through gender boundaries with less resistance, questions, or doubt about their belonging than the bodies that disturb the norm (Ahmed, 2012; Puwar, 2004).

Ahmed further explains how a minority position can make such cultural barriers and norms appear as solid as a brick wall (2012). In her studies of diversity work at universities, she presents the metaphor of the "institutional wall", reflecting the cultural barriers that make mobility in white and male academic institutions more challenging for identities that do not conform to these characteristics. The institutional wall, however, is mainly visible for those of whom they represent a barrier:

> When a category allows us to pass into the world, we might not notice that we inhabit that category. When we are stopped or held up by how we inhabit what we inhabit, then the terms of habitation are revealed to us. (Ahmed, 2012, p. 175)

Coming up against a similar cultural wall, space invaders "endure a burden of doubt, a burden of representation, infantilization and supersurveillance", Puwar explains (2004, p. 11). While these experiences can have a notable negative effect on their participation, space invaders are not simply passive bystanders. As outsiders entering the inside they *disturb* the norm and thereby reveal the social construction of the space by bringing into sight "what has been able to pass as the invisible, unmarked and undeclared somatic norm" (Puwar, 2004, p. 8). Entering spaces not intended for us produces particular ways of *experiencing* the space, and holds a potential for generating new knowledge (Ahmed, 2016, pp. 9–10) about how "marginality and privileges" are entangled in a "web of relations, bodies and space" (Puwar, n.d.). The strength of the space invader metaphor is that it captures the women's experiences not only of navigating a masculine culture and gender stereotypes, but also how their physical appearance as women affects their sense of belonging. Most of the co-constructions of gender and IT that have been documented through research remain more negative for women than for men (Cheryan et al., 2015; Ensmenger, 2012). It is still less obvious that women will pass the "if-can-test" of being categorized as IT experts (Master et al., 2016; West & Zimmerman, 1987). For some women, this might result in a wish of not drawing attention to themselves, to go under the radar (Ahmed, 2012), which adds to the challenge of, for instance, recognizing female role models in technology.

In the following chapters I will explore how women in the gender-egalitarian culture of Norway also have to deal with the challenge of fitting into masculine spaces of technology. The theories of doing gender and of resistance experienced by identities that do not fully fit the norm highlight how girls and women can experience the journey into IT differently from boys and young men, as some of the barriers are only visible to those who do not fit. The space invader metaphor is useful for understanding how women negotiate their entrance into and belonging in IT through other concepts than the most masculine images of IT. The concept of space invaders highlighting the negotiation between insider and outsider positions raises questions of how inclusion and exclusion are enacted in everyday life. In the analysis this will support our understanding of how the women find ways of establishing their sense of belonging while simultaneously challenging the masculine norm.

While the theories above point to women's encounters of cultural narratives of IT as gendered, another strand of research into women's under-representation in masculine fields of STEM, has developed a focus on how women make decisions based on their self-perception in the field. These have roots in theories from the 1970s of how women's self-efficacy in a field affect their study choice (Bandura, 1977). With roots in educational psychology, Eccles and colleagues have developed the expectancy–value theory emphasizing that the individual's expectation of her own abilities and the value she associates with the tasks are vital factors guiding study choice (Eccles, 2009; Sáinz & Eccles, 2012). Research on the use of technology has suggested that women have less self-efficacy in use of digital technologies than men (Barbieri et al., 2020), also identified in a Nordic context (Rohatgi et al., 2016). Elaborating on this theory, Master and

Meltzoff include the layer of stereotypes, which have an effect on how individuals establish a sense of belonging in certain fields (Master & Meltzoff, 2020). The stereotype threat theory predicts that the degree to which a person sees themself as fitting (or not) with stereotypes of a field, can affect their performance in that field (Steele & Aronson, 1997). The categorization of IT as a field "outside the female domain" also challenges young women's ability to identify female role models in the field (Corneliussen et al., 2019; Trauth & Quesenberry, 2007). Research has, however, identified that role model interventions can have a positive effect on women's expectations to succeed and aspirations to participate in male-dominated STEM disciplines (González-Pérez et al., 2020). Although the forthcoming analysis is rooted in a framework of social science rather than in psychology, the associations between the individual's self-efficacy and sense of belonging can support our understanding of women's experiences. While female role models are important tools for making young women identify themselves in fields of IT, the next chapters will illustrate how the women suggest alternative ways of identifying interest, self-efficacy, and a sense of belonging. The analysis suggests that we need a wider understanding of these concepts in relation to IT to understand women's strategies for entering fields of IT.

"Change is never easy but having a direction makes it possible", Holtzblatt and Marsden claim (2022). The theories engaged here will contribute to developing our understanding of how change can be achieved, as we are moving from the co-construction of gender and technology in a masculine image to a reconstruction that includes a wider notion of IT that challenges stereotypes not only of IT experts, but also of motivation and interest leading to a career in IT.

Empirical Material

The main empirical material analysed for this book involves interviews with women who study, or work with, IT. Thus, the empirical data are especially suitable for exploring how women navigate the gendered landscape of IT and for identifying the mechanisms and factors working to *support* women to choose a career in IT. The data material was collected through three consecutive research projects motivating and building on each other in the period 2018 to 2021.[3]

The first project studied women working in the primary and secondary IT sector in Norway, finding that many of the women had not originally chosen IT as their main career path.[4] In 2018 and 2019, 28 women aged 24–56 participated in in-depth interviews guided by a professional-life narrative profile with questions about education, work, and family life. Most of the women were from Norway, and they worked with IT and digitalization in public and private sectors and in research institutions. They worked with designing and programming data systems, implementing new technology, and as managers for technology development and digitalization. All had a master's or PhD degree; however, these were not all in technology, thus illustrating that many women end up working with IT even when they had not chosen IT as their main career path (Corneliussen & Seddighi, 2022).

The second project aimed to further investigate some of the main findings of the first project, in particular how women come to study IT via different routes, here by exploring women's chronological narratives from childhood to an IT degree at university.[5] The 24 women that participated in in-depth interviews during 2020 provide the main narratives of the analysis in Chaps. 3 and 4. Fourteen of these women were bachelor students, five were master's students, while five had early research recruitment positions (PhD and Postdoc) at Norwegian universities. Their main fields were in IT disciplines in STEM faculties, including computer science or informatics, computer engineering, data science, programming, bioinformatics, and cybersecurity. The women were aged between 20 and 51 years; 20 of them were born in Norway, one came from another Nordic country and three were born in countries in Asia and the Middle East. The study involved a mixture of methods, including a survey mapping already recognized factors having an impact on women's participation in IT and a drawing of their chronological story along lines of age and interest for studying IT. This worked as support during the in-depth interview for exploring what and who had affected their decisions with questions like "What happened here?" and "Why did you change direction there?", referring to the drawing.

The third study evaluated a national initiative for recruiting girls to technology and continued the search for factors and motivations that lead women into fields of technology.[6] This included in-depth interviews with 26 young women at secondary school and university level, women acting as female role models in the campaign, and teachers and school counsellors at schools participating in the campaign. A quantitative survey with nearly 700 young women in STEM education at secondary and higher education levels further expanded the study of motivational factors leading women to a career in technology. Some of the findings from the survey are presented in Chap. 2 and the interview with women are part of the empirical base for Chap. 4, while the school representatives appear in the analysis of how schools approach the issue of gender disparity in technology in Chap. 5.

Analytical Framework

The main analytical tool for the analysis across the projects has been grounded theory method (Glaser & Strauss, [1999] 2017; Strauss & Corbin, 2008), inspired by Charmaz (2006, 2017). Grounded theory represents a flexible but also systematic scientific method aiming to construct theory which is "grounded" in the empirical data. The grounded theory method comprises and guides a research project from the initial understanding and design to the final scientific publications. The method provides guidelines and support for the data collection phase, the analytical phase of reading and labelling the transcribed interviews, and the process of writing out the analysis and findings. The analytical process can be understood as a dialogue between the researcher and the data material where the researcher is constantly asking questions to the data (Charmaz, 2006). Charmaz describes the method as a strategy where the aim is to "[s]eek data, describe observed events, answer fundamental questions about what is happening, then develop theoretical categories to understand it" (2006, p. 25). The grounded theory techniques push the analysis forward through this constant movement between reading and interpreting data, labelling data and sorting labels into categories, writing memos (notes), and finally developing the scientific text. This gives impetus to "emergent theory" developed through constantly adding information and increasing the richness of the interpretation. This should not be understood as theory that simply emerges from the data, Bryant warns, since "theories of all kind need to be understood to be 'constructed' rather than 'discovered'" (2021, p. 401).

Charmaz has been in the forefront of defining and describing constructivist grounded theory method (Bryant, 2021; Charmaz, 2017). This line of grounded theory focuses on the social constructed quality of knowledge. The constructivist grounded theory approach "places priority on the phenomena of study and sees both data and analysis as created from shared experiences and relationships with participants and other sources of data", Charmaz explains (2006, p. 130). The explorative approach of grounded theory makes the method particularly useful for studying how people create meaning and why they choose their actions in different situations (Charmaz, 2006). While the method itself encourages an open dialogue to let the data material show its patterns, this process does not happen in an empty space, but rather based on the researcher's pre-understanding. The method recognizes a theoretical sensitivity reflecting the researcher's ability to understand and think analytically about the data (Charmaz, 2006). Theoretical sensitivity relies on the researcher's experience and ability to grasp meanings and nuances of the data material. Among the important sources for theoretical sensitivity are professional experience, research literature and other studies, knowledge about theories and more. This reflects the connection between the three consecutive research projects involved here, where the categories that started to take form in the first project received more input from projects two and three. The three consecutive projects contributed to making the empirical foundation for the category development more solid with in-depth exploration as well as opening for a level of generalization for some of the findings through quantitative empirical data.

The grounded theory analysis for this book involved a careful reading of the material, coding it sentence for sentence with descriptive labels. In the next stage the labels were sorted and developed into analytical notes (memos) and some of them further into categories. The central categories of the analysis presented below were sorted into different topics, while aiming to identify the *movements* happening in the data, such as the various twists and turns in the women's chronological narratives. The analysis focused on how this involved people (supporters, family, teachers, friends, role models), reflecting elements starting or distracting the women's movements (encouragement, interest in something, that sometimes, but not always included IT), environment and infrastructure surrounding their experiences (school, university, working life), and their reflections around values, competences, and gender pointing towards experiences of insider/outsider positions and ways of producing a sense of belonging.

Grounded theory is a highly popular method; however, it is also "widely criticized, often claimed without justification, and seems to arouse particularly high levels of prejudice and misunderstanding", according to Bryant (2021). Some of the critique reflects misunderstandings such as seeing this as an inductive method, where the research starts with no or few preconceptions, which, Bryant suggests, should rather be seen as "accidents" and misinterpretations of the early period of the method (2021). In this book, grounded theory has worked as a tool for driving the analytical process forward through the analytical steps described above. The topic of gender and technology has been at the core of my research since the late 1990s, thus the analytical process also included a dialogue with previous interviews, analysis, and findings (see, for instance, Corneliussen, 2011) that have contributed to the "theoretical sensitivity" as well as the richness of knowledge (Charmaz, 2006) that contributed to this book.

STRUCTURE OF THE BOOK

Many studies have explored women's underrepresentation in fields of IT education and work by studying issues and challenges in one particular context or at one stage. This book, however, provides a different approach by concentrating on the women's chronological experiences from childhood to university. Furthermore, while many studies have aimed to explore and find solutions for increasing girls and women's participation in the conventional educational routes, the analysis here provides insights into alternative pathways leading the women to IT, and alternative motivations and interests that signals new ways of developing a more gender-inclusive technology sector. The next five chapters will, based on research literature, empirical data, analysis, and discussions, develop our knowledge about how girls and women find ways of navigating the gendered landscape of IT.

Chapter 2 contextualizes the forthcoming analysis of women's experiences and pathways to IT, by revisiting relevant research literature, mainly from western countries, about the status for girls' and women's low participation in IT. The chapter highlights knowledge about barriers women encounter on their way to a career in technology as well as factors that attract women to technology. Despite many initiatives for recruiting girls and women to fields of technology, the effects are often short-lived. Thus, the chapter engages in a discussion of how assumption of gender equality already in place in the Nordic countries can become a barrier for promoting the very same values.

Chapter 3 presents the analysis of how women have come to IT via different pathways. The chapter focuses on factors that have positively influenced women's decision of studying IT. The analysis builds on the interviews and drawings made by the 24 women who contributed their narrative of their chronological pathway from childhood until the decision to embark upon a university degree in IT. The analysis illustrates six different pathways that the women describe, each analysed in terms of the positive drivers that have made the women decide to study IT. The six pathways demonstrate that most of the women navigated the landscape of IT in very different ways from the ideas highlighted by the computer science professor, the IT manager, and the educators quoted above.

Gender remains unspoken in the analysis of women's pathways to IT in Chap. 3. Yet gender was inextricably entangled in their experiences, and Chap. 4 explores how this affected their journey. The women's narratives confirmed that gender stereotypes were still widespread, challenging their engagement. However, they also found solutions to bypass, negotiate, and challenge such stereotypes, by developing their own rhetoric for justifying their fit with and belonging in IT. In the process, the women suggested new ways of understanding IT and they claimed visibility for women in IT. The women's narratives illustrate how their perceptions of IT developed from an initial understanding defined by gender stereotypes to a more inclusive understanding that supported revised visions about IT as well as their own sense of belonging there.

Recognizing that women do not operate in isolation when navigating the landscape of IT, Chap. 5 shifts perspective to schools, which can play a significant role in making women familiar with IT. Here we explore how representatives from 12 Norwegian secondary schools consider their role in encouraging girls and women to study IT. Although gender equality represents a vital value in Norwegian schools, the school representatives illustrate diverging views in terms of what gender equality means and how to achieve it. The chapter involves a discussion of how certain attitudes, such as a distrust to women's interest in IT, affect the schools' ability to become a supportive arena for recruiting women to IT.

Chapter 6 sums up the lessons learnt from the women's narratives and their space invader experiences as well as the schools' mixed and rather vague responses to issues of gender disparity in IT. This involves a discussion of the consequences of the new knowledge for research, potential supporters for women, policymakers, and not the least women, in particular those who had never thought about choosing IT at school. Here this raises questions about the mismatch between a strong gender equality norm and a continuous low participation of women in most fields of technology—the Nordic gender equality paradox. Contrary to the oft-repeated assumption of women's career preferences holding the explanatory force to this paradox, the evidence here suggests that a myth of gender equality already in place rather makes the gender equality norm "non-performative" (Ahmed, 2012), and thus in some ways it acts as a barrier to the very values that it names.

The women's experiences of barriers, as well as their solutions to work around them, will over the next chapters be developed into a more thorough and coherent understanding of how young women navigate the increasingly digitalized world. While the concept of co-constructing gender and technology with an emphasis on the mutual shaping of the two has represented an important approach to understanding women's underrepresentation in IT (Bray, 2007; Cockburn, 1992; Cockburn & Ormrod, 1993; Landström, 2007; Wajcman, 2010), it is time to move further and explore how women's own strategies for entering fields of IT are contributing to specific ways of *reconstructing* the relationship between gender and technology. This perspective will be further developed in the final chapter of the book, with potential users of this knowledge in mind, such as scholars and practitioners, educators, employers, policymakers, and anyone else interested in strategies for making IT a more gender-inclusive field.

NOTES

1. I prefer the term information technology (IT) in this context since the term information and communication technology (ICT) has often been defined as related to the use of technology rather than producing (with) technology (see Dee, H. 2021).
2. Puwar builds on Charles Mill's concept of white male bodies as a somatic norm (Puwar, 2004, p. 33).
3. The three research projects including data collection and management have been approved by the Data Protection Services at the Norwegian Centre for Research Data. All interviews were recorded and later transcribed verbatim. All informants received information about the aim of the study and signed a letter of consent. All informants have been anonymized for the analysis.
4. This study was part of the *Nordic Centre of Excellence, Nordwit* focusing on women in tech-driven careers, funded by *Nordforsk* between 2017 and 2022. The study included several sub-projects with interviews with men and women working with technology in Norway, Sweden, and Finland.
5. This project was an assignment for the *Norwegian Centre for STEM Recruitment* and the author was responsible for the entire project. Two assistants helped with seven of the interviews.

6. This project was an assignment for *The Norwegian Directorate for Children, Youth and Family Affairs* (Bufdir) with the aim of evaluating the national campaign *Girls and technology* and to identify what motivates young women's study choices related to technology and other science, technology, engineering, and mathematics (STEM) fields.

CHAPTER 2

The Unsolved Mystery of the Gender Imbalance in IT

Abstract Women are underrepresented in information technology (IT) education and work across the western world. This chapter contextualizes the topic of the book by revisiting research literature about girls' and women's participation in IT. Among the widely recognized barriers are gender stereotypes and gender structures in IT education and work. The chapter further reviews studies into motivational factors as well as research investigating women entering IT through non-traditional training grounds. Finally, the question of why the situation has not improved faster in the Nordic countries is discussed in light of the metaphor of a Nordic gender equality paradox. This reflects a gap between theory and practice and a myth of gender equality already in place that reduces efforts to address gender inequality in technology.

Keywords A western challenge • Alternative training arenas • Gender stereotypes of IT • Motivation for studying IT • Recruiting women to technology • Women's underrepresentation in IT

INTRODUCTION: A WESTERN CHALLENGE

Women's underrepresentation in most fields of IT has been a topic for researchers across the western world for decades (Arnold et al., 2021; Cohoon & Aspray, 2006; Corneliussen, 2011; Haraway, 1991; Hayes,

© The Author(s) 2024
H. G. Corneliussen, *Reconstructions of Gender and Information Technology*, https://doi.org/10.1007/978-981-99-5187-1_2

2010; Holtzblatt & Marsden, 2022; Wajcman, 1991). It is a complex and multifaceted challenge that has changed over time and place (Charles & Thébaud, 2018; Cohoon & Aspray, 2006; Misa, 2010b). There are, however, also similarities, in particular a notable slow improvement, stagnation as well as setbacks (Branch, 2016), showing that although context and situation might be specific to national cultures, most western countries share the challenge of producing effective and long-lasting improvements to the situation.

Research into the underrepresentation of women in IT covers different topics, some relating to challenges of recruiting, others relating to women's challenges in a male-dominated work culture of IT and a large number of women leaving the sector (Branch, 2016; McKinney et al., 2008; Pantic & Clarke-Midura, 2019). Though there are overlapping features, these different perspectives involve different challenges and raise different questions. There has been little focus on the issue of women leaving the tech sector in Norway. Recent numbers from *Statistics Norway* shows that women who complete a degree in the technology are as likely as men with similar degrees to work in the field (Foss, 2020). Recruitment thus appears to be the most pressing challenge for a more gender-inclusive tech sector in Norway. This book will explore this issue with the aim of learning more about what brings women to higher education in IT.

This chapter will revisit the knowledge landscape of girls and women's pathways to IT in the western world and with a special focus on Europe and Norway. It will do so by looking at research into barriers and women's motivations for participation in IT, the relevance of support from family and educators, before considering explanations for the slow improvement in this field.

Numbers Talking

IT employment in Europe has increased in the past decade, growing notably faster than employment in general (Eurostat, 2021c). The gender gap, however, remains large. In 2020, the average participation of women in IT work in Europe was 18%, ranging from 10% in Czechia to 28% in Bulgaria, with Norway at 19% (Eurostat, 2021a). Different from the situation in Norway, women leaving the field during their first years of working in IT is a challenge reported from the US as well as the European Union (EU) (Griffiths & Moore, 2010; Holtzblatt & Marsden, 2022; European Commission, 2013). The annual productivity loss for the

European economy due to women leaving their digital jobs was stipulated to more than 16 Billion Euros in 2018 (Quirós et al., 2018).

Education in Norway is, in general, state-supported, thus reducing inequality due to economic accessibility of education. Gender norms and stereotypes, however, still have a large effect on youths' educational choices (NOU, 2019: 19), and women's participation in certain fields of technology education are lower in Norway than the OECD average (OECD, 2021). Closing the gender gap in technology and the other science, technology, engineering and mathematics (STEM) fields will not only increase employment and empowerment of women in Europe, but also foster economic growth (The European Institute for Gender Equality, 2017). There has been a positive development for women's participation in information technology disciplines in higher education in Norway, with an increase from 17% in 2012 to 29% in 2021, as illustrated by Fig. 2.1. This is a positive trend, however, there are still large variations between universities and disciplines, ranging from less than 10% in disciplines concentrated on programming, to more than 50% in disciplines focusing on design, use and interaction (The Norwegian Universities and Colleges

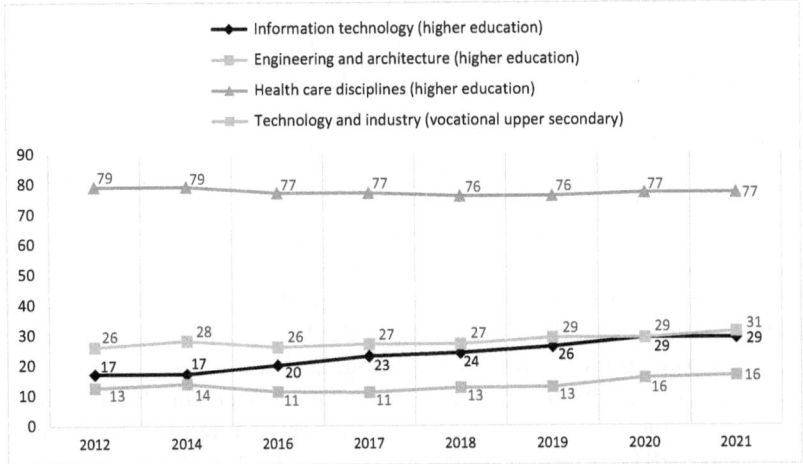

Fig. 2.1 Female applicants in percentage to higher education in technology and health care, and vocational courses of technology and industry from 2012 to 2021. (Source: The Norwegian Universities and Colleges Admission Service and The Norwegian Directorate for Education and Training)

Admission Service, 2022). Furthermore, the increase in women's participation at university level has not yet produced a notable increase in women's participating in IT jobs in Norway, different from the development in some of the other European countries (Palmer, 2021; Simonsen & Corneliussen, 2020).

The proportion of women in technology disciplines in vocational education in upper secondary level is even lower and show less signs of improvements. A recent evaluation of the Norwegian government's strategy for increasing participation of children and youth in STEM also found discouraging results for both boys and girls. The evaluation showed a particular low motivation among girls for choosing technology subjects such as programming, with only 7% girls against 31% boys in 2020 (NIFU, 2021). Education in health care experience the opposite challenge, with weak recruitment of boys. Figure 2.1 includes health care in higher education and thus visualizes the large gender differences in educational choices in the Norwegian educational sector. As the figure illustrates, there have been some improvements in this field, however, the main pattern of a notable horizontal gender segregation is still unmistakably clear.

Barriers for Women's Participation in IT

Women's underrepresentation in IT is not only notable in the statistics but is also reflected in narratives describing IT as a "world without women" (Sørensen, 2011, p. 45) and technology as "gender inauthentic" to women (Faulkner, 2009, p. 172), opposed to narratives of "men's love affair with technology" (Oldenziel, 1999, p. 9; Hacker, 1989; Kleif & Faulkner, 2003). Gender stereotypes about IT as well as gender structures in IT education and work are identified as some of the most challenging barriers for girls and women's engagement in IT (Chavatzia, 2017; Cheryan et al., 2015; Cohoon & Aspray, 2006; Frieze & Quesenberry, 2015, 2019; Turkle, 1988; Wajcman, 2004; Watts, 2009). Despite variations between countries, it is widely recognized across the western world that gender cultures, norms, and stereotypes affect and limit youths' choices of education in ways that reproduce the gender imbalance in IT (Chavatzia, 2017; Frieze & Quesenberry, 2019; OECD, 2016). This is also the case in Norway, as documented in a recent governmental report concluding that youth in Norway are highly affected by gender norms and stereotypes when choosing their education and career paths (NOU, 2019: 19). Gender stereotypes defining IT as a masculine field make women less likely to identify with IT, while men are more likely than women to believe that they "match the cultural image of successful tech workers" (Wynn & Correll, 2017, p. 5). A study among OECD countries reflects this with only 0.5% of girls compared to 5% boys aiming for a career in IT (Borgonovi et al., 2018).

Gender stereotypes are responsible for many different types of barriers for women's participation in IT (Master et al., 2016; Master & Meltzoff, 2020; Yates & Plagnol, 2022). It has been noted that youth in general have limited knowledge of and understanding for what IT education and IT work represents (Grover et al., 2014; Jethwani et al., 2016). A recent Swedish report shows a discouraging low and sinking level of knowledge about IT work among young women (Insight Intelligence, 2022). Lack of insight gives youth unrealistic expectations and opens up the opportunity for stereotypes to dominate the youths' perception of IT (Spieler et al., 2019). This also suggests that lack of insights about IT have a more negative effect on young women than it has on young men, who can rather lean on positive stereotypes predicting favourable beliefs about men's relationship with IT (Czopp et al., 2015).

A well-studied field is the harmful effect that negative stereotypes about a group can have on self-confidence and trust in their own abilities to perform key tasks, such as predicted in the theory of stereotype threat (Steele & Aronson, 1997). The expectancy–value theory (Eccles, 2009) suggests that individuals' study motivation is influenced by their expectations of success as well as the value and rewards they identify with a field or discipline. This has contributed to studies exploring how a lack of suitable images, role models, and support combined with a low expectation of success can produce significant barriers to girls' and women's participation in fields of technology (González-Pérez et al., 2020).

A large bulk of studies have documented that gender stereotypes and images of IT as masculine challenge women's entries into IT. This has, for instance, been illustrated by Lewis and colleagues, who found that women who associate computer science (CS) with "traits of singular focus, asocialness, competition and maleness" interpreted this as an indication that "men are innately more talented" than women in CS (Lewis et al., 2016, p. 23). Such images make women doubt that they can compete with fellow students and colleagues (Margolis & Fisher, 2002; Yates & Plagnol, 2022).

This also echoes stories of the dawning computing industry in the late 1960s, aiming to define a programmer so that they could recruit the best ones. Emphasizing certain types of personality traits, the result was a description of programmers as "antisocial, mathematically inclined males". This contributed to a self-fulfilling circle of recruiting what they named, and thus the overrepresentation of "antisocial, mathematically inclined males" worked to confirm that the definition was appropriate (Ensmenger, 2012, pp. 78–79). Research from the US suggests that the impression of programming as an innate ability that is mostly found among men, is still strong, and is stronger in computing than in other disciplines (Becker, 2021; Guzdial, 2015). Such assumptions produce notable barriers for increasing diversity in fields of IT, also identified in the UK by Yates and Plagnol. Their study showed that many women studying computing felt that their technical competence was questioned, and the women shared an experience of "feeling stupid" compared to their fellow male students (2022). One consequence was that women tended to back out from job reviews when they realized that their coding abilities were put on the test, because they assumed they would lose out to men they imagined had been coding since they were "in diapers" (Yates & Plagnol, 2022). Women facing male co-students that seem to confirm this gender stereotype tend to reconsider their own position by questioning whether they have the right or enough interest (Margolis & Fisher, 2002).

According to social cognitive theory, gender stereotypes might affect women's self-efficacy in male-dominated fields such as IT (Bandura, 1977). This points to the importance of providing female role models in IT, which is often a core activity in recruitment initiatives targeting women (Dasgupta, 2011; Lang et al., 2020). However, identifying female role models in technology can be problematic for young women (Arnold et al., 2021; Corneliussen et al., 2019; Thomas & Allen, 2006). One of the informants in our study of women in IT work explained that she didn't know of any female role models, "because often there haven't been anyone before us, in a way" (Corneliussen et al., 2019, p. 383). Such claims appear to reflect not only the scarce number of women in IT, but also the masculine coding of the field which seems to give women a feeling of being the only or "first" woman with a career in IT (2019).

A Nordic Gender Equality Paradox

Some researchers have posed the question: Why do we find this situation also in the gender-egalitarian Nordic countries? (Charles & Bradley, 2006; Chow & Charles, 2019; Stoet & Geary, 2018). The pattern of gender segregation, including women's low participation in IT, appears to be in conflict with the image of the Nordic countries as "superpowers" of gender equality (Larsen et al., 2021, p. 2; Ellingsæter, 2014; Sund, 2015). This gender equality paradox takes on a Nordic profile in international comparisons highlighting gender inequalities which appear as more extreme in highly gender-egalitarian and affluent countries (Chow & Charles, 2019; Minelgaite et al., 2020; Stoet & Geary, 2018). The paradox is shaped by a failed expectation of gender equality in one field leading to gender equality in other fields (Ellingsæter, 2014, p. 101). Authors behind international studies have suggested that the paradox is a result of countries with a high level of gender equality and low "life risks" opening for individuals to choose education and career according to "individual interests and academic strengths" (Stoet & Geary, 2018, p. 582). This suggests that the gender gap in fields of IT in gender-egalitarian countries such as Norway is a result of women having *less interest* for these fields, or even less career ambitions, and less academic strengths in IT and related subjects. Other studies have challenged this view, pointing out that the question of interest is not simply an individual isolated preference, but needs to be considered in the context of gender stereotypes and a masculine culture of computing (Blum et al., 2007; Cheryan et al., 2017; Corneliussen, 2021a; Yates & Plagnol, 2022). Furthermore, women having a lower career ambition than men is not supported by the fact that more women than men apply for higher education in general, and that they also outnumber men in some of the educations traditionally associated with high-prestige professions such as medicine, psychology, and law (The Norwegian Universities and Colleges Admission Service, 2022). Thus, the paradox rather seems to reflect a gap between theory and practice (Minelgaite et al., 2020), or between a "myth" about gender equality being already in place despite continuous inequality and a culture that does not provide necessary support to achieve gender equality in practice (Holtzblatt & Marsden, 2022; Martinsson & Griffin, 2016). This has caused many contradictions experienced by women in the Nordic countries, where on the one side there is a strong public gender equality rhetoric (Griffin, 2022) emphasizing that women are wanted and needed in fields of technology, while on the other side women still experience barriers for participating in IT contexts.

WHAT BRINGS WOMEN TO IT?

We know a lot about barriers for women's participation in IT, but what makes girls and women consider IT a relevant career choice? Cheryan and colleagues have suggested that there are three main areas that can benefit from more gender-inclusive images to make IT more attractive to girls and women. This involves altering the stereotypes and "broadening the representation of the people who do this work, the work itself, and the environments in which it occurs" (2015, p. 1). Providing girls with an early opportunity for becoming acquainted with computers is one strategy for providing relevant insight into what it means to study and work with IT. Early exposure to computers and concepts of computer science can provide girls with knowledge and familiarity of IT that can reduce effects from negative gender stereotypes. Starting early is important because negative attitudes to IT tend to grow stronger during the teenage years (Cheryan et al., 2013), for instance involving images of IT as boring and unsuitable for girls (Armoni & Gal-Ezer, 2014; DiSalvo et al., 2014; Prottsman, 2014). Research has suggested that such an approach can support girls' interest in studying IT later (Jones & Hite, 2020; Yates & Plagnol, 2022).

Boys have, in general, been more exposed to technology at early stages than girls (Barker & Aspray, 2006; Gerson et al., 2022). Thus, the early exposure strategy partly reflects the typical success stories associated with boys' interest in studying IT. Some studies have shown that early exposure to technology does not necessarily increase girls' interest in studying IT (Vrieler et al., 2020), while for women who have decided to study IT, it can appear vital. For some of the women interviewed in Yates and Plagnol's study, thinking of studying computer science without having had early exposure to computers, in particular to coding, was considered a risk (2022). With more exposure to IT, in particular to programming, girls develop their self-efficacy in relation to technology as well as their career orientation towards IT (Aivaloglou & Hermans, 2019). Not the least, this has a positive effect on women's fear that other students, whereof the majority tend to be men, are more skilled because they have been playing around with technology (Master et al., 2017; Yates & Plagnol, 2022).

While a surge of code clubs in the 2010s invited children to learn about coding as a strategy for recruiting future dedicated IT students, limited focus on recruiting girls to such arenas risks reproducing gender stereotypes and the gender imbalance in IT (Corneliussen & Prøitz, 2016).

Furthermore, while studies of early exposure suggest that such activities can represent important insights into IT, the variation and partly contradictory conclusions also indicate that exposure alone is not enough to challenge the gender coding of IT as masculine. One reason might be that many of the elements that have been considered to provide motivation for and interest in studying IT have developed from narratives about boys' playful relationship with IT and are already associated with a masculine relationship with technology. If the reflection of IT as a male field is a strong one, the same activities that can have positive effects on boys, can have rather discouraging effects on girls' and women's aspirations for pursuing IT. Girls need to see computing as a subject where also they belong, as Dee emphasizes (2021, p. 45).

This leads to the question of what IT represents, which also needs to offer something that interests women. In today's Norwegian society and the level of digitalization experienced here, the answer is that IT can be a vital instrument for nearly anything. One challenge is that many studies of women's underrepresentation in IT set out to measure aspiration to participate in terms of attitude and interest in the context of rather traditional computing activities and arenas (Eccles, 2009; Master & Meltzoff, 2020; Sáinz & Eccles, 2012). Questions about interest often reflect assumptions about boys' gateways to IT building on interest raised through leisure activities and in private spaces. Boys' interest in competition and gaming has been identified as giving advantages for their participation in IT education and work, for instance by Sevin and DeCamp, who suggest "game play is statistically significant as a predictor of confidence and interest" for studying computer science (2016, p. 1). They suggest that video games as "a recreational technology" help develop the players' abilities to navigate the landscape of "non-recreational technology" and to increase their interest in technology in general (Sevin & Decamp, 2016, p. 14). This supports assumptions that gamers acquire some types of insights in IT that helps them against the identified lack of knowledge of IT among youth (Jethwani et al., 2016). Social cognitive theories developed around concepts of individuals' trust in their own abilities to master and succeed in a field in close operation with a sense of belonging (Eccles, 2011; Eccles & Wigfield, 2002; Sáinz & Eccles, 2012, see Chap. 1), can explain why, for instance, boys' experiences with computer games can translate into an interest for studying IT. Experimental research supporting this has, for instance, found that when IT disciplines and environments are presented in a gender-neutral way rather than reflecting gender stereotypes, girls and women express more interest in the field (Cheryan et al., 2015; Master et al., 2016). Women also react positive to lecturers of both genders who do not display stereotypical notions of IT, while men in the same contexts do not respond much different to this (Master et al., 2014). Thus, when efforts are taken to reduce, erase, or transform negative gender stereotypes, girls' interest and aspiration for participating increases, while boy's aspiration seems less affected (Master et al., 2016).

A survey with nearly 700 young women (age 16 to 36+) explored what had motivated them to consider or deciding to study technology at upper secondary level or in higher education (ISCED 3–6) (Corneliussen, Seddighi, Simonsen, et al., 2021; Corneliussen, Seddighi, Urbaniak-Brekke, et al., 2021). As shown in Fig. 2.2, most of the women agreed that exciting job opportunities were important, closely followed by the importance of technology knowledge, good salary, and technology as important for solving societal challenges. These numbers point to motivational factors that are rather similar between men and women; however, it also identifies women's emphasis on societal aspects as a motivation for working with technology. The least important motivational factors were related to leisure activities and computer games, documenting that some of the triggers for choosing a technology career associated with boys

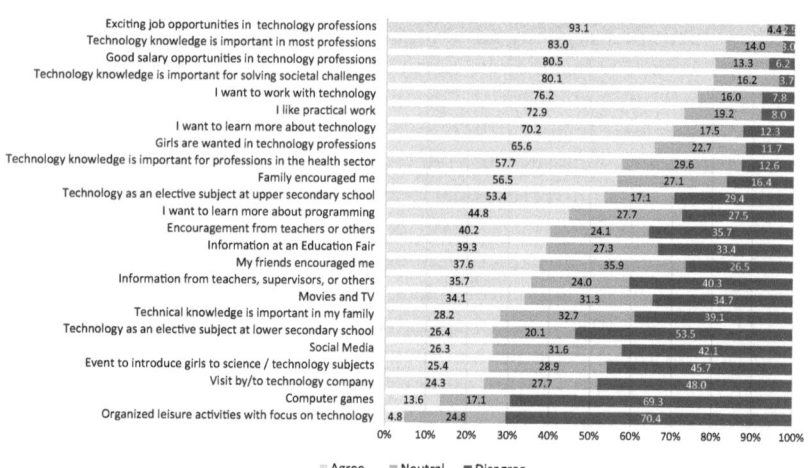

Fig. 2.2 Motivational factors for choosing technology among women (percentage, N689). (Source: Corneliussen et al., 2021)

(Sevin & Decamp, 2016; Yates & Plagnol, 2022) are of little importance for most girls and young women.

The survey also provides information about how the women experienced support from friends, family, and school. Technology at upper secondary school and encouragement from family were important motivators for more than 50% of the respondents (see Fig. 2.2). The high number of women identifying IT classes at upper secondary school as an important motivation for studying IT is notable compared to the relatively small proportion of women participating in these classes on a yearly basis (The Norwegian Directorate for Education and Training, 2023). Another Nordic study found that family was the most important motivation for early decisions to study technology (Engström, 2018). Whether the motivation behind choosing IT at secondary school stems from family background cannot be identified in this survey. However, the high proportion of respondents identifying the IT classes as motivation for further studies in IT suggests that they have an important role in preparing women for higher IT education. Apart from the technology classes, Fig. 2.2 shows that school had a modest role in motivating young women to consider a career in technology. Only 1 in 4 had been motivated by recruitment initiatives.

Considering the question of who the IT workers are, can also help to identify strategies for motivating women. The high importance of family with a background in technology illustrated in both the above-mentioned Nordic studies (Corneliussen, Seddighi, Simonsen, et al., 2021; Corneliussen, Seddighi, Urbaniak-Brekke, et al., 2021; Engström, 2018) also points to the value of identifying and recognizing IT professionals. It has been challenging for women to identify female role models in technology. The reason is not only that men outnumber women in tech work, but also that cultural images and stereotypes picture the IT expert as a man, thus "doing IT while doing femininity" appears like a contradiction (Corneliussen et al., 2019; Nentwich & Kelan, 2014).

Educational choices are often assumed to rely on rational choices (Reisel et al., 2019), assumptions about one's own abilities or—in particular, when talking about technology—to be inextricably dependent on interest (Sáinz & Eccles, 2012), often developed in spaces and places of leisure time (Sevin & Decamp, 2016). Many studies have, however, shown that this also goes hand in hand with the ability to associate with the main characters recognized in the field. Female role models can increase girls' expectations to succeed in a male-dominated field such as IT

(González-Pérez et al., 2020). Seeing someone they can associate with is important for girls' ability to imagine themselves in a future career in IT (Lang et al., 2020; NOU, 2019: 19). Students tend to choose careers that include features that they associate with themselves (Eccles, 2009). Role models, in particular women (European Commission, 2013; Stout et al., 2011) and role models that do not embody gender stereotypes of IT, have been identified as important for making girls' able to imagine themselves in the field (Master et al., 2014). Within the tradition of expectancy–value theory, Gonzales-Peres and colleagues found that exposing girls to female role models with a successful career in IT or, more broadly, STEM fields, increased girls' expectations of success and decreased the negative effect of gender role stereotypes (González-Pérez et al., 2020). Although role modelling involves complex issues, as discussed above, also the interviews analysed in this book suggests that providing girls and women with images that they can relate to can have a decisive effect on study choices.

Effects of Recruitment Measures

Several decades of research into the underrepresentation of girls and women in fields of technology have created important knowledge about what the challenges are as well as producing suggestions for how to make changes. Putting such solutions to work, however, is an understudied field. Actions to recruit girls and women to technology are few, mostly local, and most have short-lived effects (Lagesen et al., 2021). Few recruitment initiatives have been evaluated, thus it remains to identify precise effects (Reisel et al., 2019). One exception to both is the national campaign, *Girls and Technology*, which was evaluated by the author and colleagues in 2020 (Corneliussen, Seddighi, Simonsen, et al., 2021; Corneliussen, Seddighi, Urbaniak-Brekke, et al., 2021). This state-financed campaign has since 2016 been organized by the largest organization for employers in Norway, NHO, together with partners. The aim of the campaign is to encourage girls and women to choose technology educations. The main event of the campaign is a national tour of female role models. Across the country, girls from lower and upper secondary school are invited to an "inspiration day" with motivational presentations of technology education and professions, promoted through female role models. The evaluation showed that the emphasis on women studying and working with technology was highly appreciated among the girls and young women, because it represented an alternative to the more widespread perception of technology as "something boys do together with their fathers", one of the girls explained (Corneliussen, Seddighi, Simonsen, et al., 2021, p. 23). The girls-only quality of the events created a community of girls that included an acceptance of girls also being interested in and playing with technology (Corneliussen, Seddighi, Simonsen, et al., 2021).

The evaluation of the campaign confirmed that gender-stereotypical attitudes could have a negative impact on girls' aspirations to study technology, and, more importantly, meeting non-stereotypical female role models made girls more open for choosing technology. Although it can be difficult to isolate and identify the precise reasons for deciding to study technology, the evaluation showed that nearly all the women who participated in one or more events before choosing educational direction, reported that the campaign had an impact on their decision to study technology (Corneliussen, Seddighi, Simonsen, et al., 2021). Based on interviews with 26 women and a survey with nearly 700 respondents, the evaluation report points to two important effects of such recruitment initiatives: firstly, for girls who do not find support for learning about technology from family or school, the recruitment initiative compensated for this. The focus on female role models presenting new and exciting technology could have a very dramatic effect on some of these girls who had not imagined themselves studying technology before. It worked almost like a "rocket", transporting them straight into a new landscape where technology could appear a natural choice not only for boys, but also for girls. This had made some of them leave behind their former study choice of a more gender-traditional discipline (e.g. social science or health care) to explore further the possibilities in IT. Secondly, for those girls who already were interested, seeing female role models was adding to their former experience and could be vital for their ability to imagining themselves working with technology. This effect is less dramatic, but equally important for supporting the young women towards the final decision to study technology (Corneliussen, Seddighi, Simonsen, et al., 2021, p. 6).

The *add-on* as well as the *rocket* effect suggest that girls need insight into what technology is, what it is used for, and, not least, to see also that women can be experts to make them consider technology as a suitable, interesting, and welcoming field. Reflecting Puwar's concept of "space invader" (Puwar, 2004), the recruitment events made the young women feel less like an outsider to a masculine space of technology. Instead of meeting technology as a "gender inauthentic" educational choice (Faulkner, 2009), the girls were invited to experience an insider position in a gender-inclusive space in which women too were identified as experts.

ALTERNATIVE TRAINING ARENAS

As Fig. 2.1 illustrates, most young women in Norway do not consider IT a relevant study choice at the time when they were moving from upper secondary to tertiary education or higher education. Furthermore, the evaluation of the *Girls and Technology* campaign suggested that the schools were only partly successful as motivator for girls to study IT. The transition from lower to upper secondary school, and from secondary to university, are identified as contributing to gender differences in educational choices (Reisel et al., 2019). Alternative training arenas, such as out of school initiatives, are less often involved in research aiming to understand women's participation in IT, and this topic is often limited to the context of the younger children and teenagers (Corneliussen & Prøitz, 2016; Vrieler et al., 2020). A limited number of studies of women finding alternative training grounds identify boot camps, apprenticeships, and other non-traditional arenas for learning about programming and IT as important sites for recruiting women who had not been recruited through school (Lyon & Green, 2020; Smith et al., 2020). Some of these alternative routes to IT competence attract a higher percentage of women than tertiary education and university degrees (Lyon & Green, 2020; Seibel & Veilleux, 2019; Smith et al., 2020), and appear attractive for women who had not aimed for an IT education as undergraduates (Seibel & Veilleux, 2019).

In a Nordic study of women working in the primary and secondary IT sector, the 28 women interviewed were engaged in designing and programming as well as implementing new information technology, and some working as managers for processes of IT innovation and digitalization (Corneliussen & Seddighi, 2022). Less than half the group had pursued a "conventional" educational route of choosing math and sciences at upper secondary school and then moving directly to an IT program at university. A larger group had pursued alternative routes to IT work, mainly triggered by new requirements for IT skills and digital competence across different fields and occupations. Many of the women had started within a non-tech and rather gender-traditional education such as nursing, economy, and social science, before changing direction towards technology. Some returned to university and others engaged in work-based upskilling in IT, while some of the women had been engaged in core processes of digitalization because a wide set of competences beyond the tech disciplines were needed in processes of digitalization (Ekeland et al., 2015).

This study thus further illustrates that many women who had not imagined working with technology had still made choices that led to a career in technology. One of the women described this as a *natural progression* from a master's degree in chemistry to a PhD in cybernetics. In the interview we asked why she had made these choices: "Well, in fact I chose chemistry. When I finished (upper secondary school) I didn't even know what cybernetics was. And I am not sure that I would have chosen it even if I had known" (Dani, quoted in Corneliussen & Seddighi, 2022, p. 66). The barriers that many women experience when approaching tech education during their teens, might not appear equally daunting when they move into tech via less conventional routes, such as Dani's natural progression. Although many of the women in this study had developed a high degree of involvement and central positions in processes of digitalization with a vital importance for society, many of them will remain invisible in the public statistics identifying individuals who are moving through the more traditional educational routes to tech work.

STEREOTYPES AFFECTING SUPPORTERS

Most research exploring women's underrepresentation in technology focus on girls' and women's experiences and attitudes, in line with the literature reviewed above. A large bulk of studies have explored the role of family and teachers for girls' aspirations to study IT (Gerson et al., 2022; Jacobs et al., 2017; Jethwani et al., 2016; Master et al., 2014). Despite some variations in the results, there is convincing evidence that parents and family members can have an important impact on girls' aspiration to study IT, as role models and by making a career in technology appear acceptable for girls (Ardies et al., 2021; Engström, 2018; Holtzblatt & Marsden, 2022; Tandrayen-Ragoobur & Gokulsing, 2021). The impact of parents' background was also significant in the survey among women of motivational factors for studying technology. Here 70% of women in higher IT education had parents with university education and 70% had one or more family members working in technology or one of the other STEM fields. Teachers, lecturers and other "influential individuals", in particular "male champions", can also have important roles in supporting women's entry into IT degrees (Yates & Plagnol, 2022; Master et al., 2014) or for dealing with barriers in the educational system (Trauth & Connolly, 2021). However, Nordic research has found that schools play a minor role compared to family with relevant education as motivation for girls' choice of technology education (Corneliussen, Seddighi, Simonsen, et al., 2021; Engström, 2018).

Considering the current insight about girls' and women's need of knowledge and encouragement to start thinking of IT as relevant, family and school appear as important actors for supporting this. A less-studied effect is, however, how parents, educators, and other potential supporters for girls' participation in IT see their own role and responsibility in this scenario, and how they are affected by the cultural narratives and stereotypes that threaten girls' self-image in IT. The gender stereotypes that work negatively on women's aspirations to study IT also affect this group. Documenting this, a Danish study found that 70% of parents assumed that boys were more interested in IT than girls, and only 1% of the parents imagined girls to be more interested in technology than boys (Tænketanken DEA, 2019). Gender stereotypes work as a filter for these groups to judge girls' and women's relationship with technology (Vekiri, 2013; Yates & Plagnol, 2022). Stereotypes also affect to which degree girls are actively encouraged and motivated to participate in technology training and education. Seeing the gender imbalance in IT as the *normal state of affairs* results in less effort put into recruiting girls and young women to technology arenas (Corneliussen & Prøitz, 2016). Reflecting similar stereotypes, several studies have documented that women are judged as less competent than men (Holtzblatt & Marsden, 2022; Moss-Racusin et al., 2018), and they are less likely to experience positive encouragement from supervisors in tech companies (Wynn & Correll, 2017). A study of attitudes among representatives from the Norwegian IT sector found a widespread doubt about women's interest in technology. Here, this transformed into a doubt about women's competence. Men have a "big advantage" before women in IT education, one of the IT experts suggested, because they had already developed their IT skills through their early playing and fiddling with computers (Corneliussen & Seddighi, 2019, p. 281). "Do we really need more women in IT? We can't force women to like IT. What if they are more interested in other stuff?", one of the IT experts asked (Corneliussen & Seddighi, 2019, p. 281). These attitudes had consequences for recruitment, and the IT sector representatives had not considered themselves neither responsible for, nor able to change women's underrepresentation in IT.

Thus, the barriers and challenges that girls and women face are not only related to their self-perception, ability belief, or whether they can associate with images of IT expert. It is also shaped by other actors' and potential supporters' gender-stereotypical assumptions about and expectations to girls' and women's relationship with technology.

CHALLENGING THE NOTION OF GENDER EQUALITY

The research literature on cultural images and gender stereotypes working as barriers for girls' and women's engagement in fields of IT, from leisure to education and professional arenas, is overwhelming. With all the knowledge we have about challenges and barriers, a pertinent question is: Why isn't the situation improving faster? Although we are not short of identifying factors working as barriers, there is little evidence that this research has produced adequate solutions to overcome the barriers, apart from local and temporary improvements (Lagesen et al., 2021). The recent evaluation of the Norwegian strategy for recruiting more girls and boys to STEM disciplines could not register improvements (NIFU, 2021). Another study of a long-term initiative to recruit women to IT disciplines at a Norwegian university found that the measures were working; however, the effect faded when the initiative ended (Lagesen et al., 2021).

There is ample evidence demonstrating that gender equality does not arise automatically, but rather requires conscious and intentional efforts over time (Devillard et al., 2016; Frieze & Quesenberry, 2015; Hunt et al., 2018; Lang et al., 2020). This also applies to the Nordic countries. However, the gendered patterns in education and working life have often been understood as a result of boys' and girls' different *interests*, rather than of gender discrimination (Snickare & Holter, 2021). The strong myth about gender equality in the Nordic countries makes it appear as if gender equality has already been achieved. Research has illustrated that instead of the gender equality norm working as a call for action, the same widely accepted norm itself can be interpreted as the solution (Corneliussen & Seddighi, 2020a). In other words, when gender equality is assumed to exist, less effort is put into producing gender equality (Ahmed, 2012). Kaiser and colleagues have through experiments demonstrated that diversity structures in a company creates "an illusion of fairness" that makes the majority group less sensitive for recognizing discrimination, while those claiming to experience discrimination risk negative reactions from the majority group (2013, p. 504). They conclude that the presence of structures aiming for diversity produce the "ironic consequence of reducing perceptions of discrimination and undermining support for those who claim to be its victims" (Kaiser et al., 2013, p. 516).

The Norwegian gender equality ideology has also been identified with a similar effect. Exploring gender equality in academia, Holter and Snickare found that men and women make many of the same career choices. Despite finding only small differences, they documented a widespread discourse in which the argument of gender differences in career choices appeared as a legitimate way of referring to men and women as *different, but equal* (Holter & Snickare, 2021). This supported a view of gender inequality in the academic setting perceived as men and women's different choice rather than being a result of a lack in gender equality (Snickare & Holter, 2021). The authors suggest this is related to the impression of Norway as one of the most gender-equal countries in the world, which is often misinterpreted as a *description* of Norway as gender-equal (Snickare & Holter, 2021, p. 24). Furthermore, seeing gender differences as a result of individual choices undermines the will to intervene based on the attitude that individual choices *neither should nor could be forced* into a different pattern (Snickare & Holter, 2021, p. 26). Furthermore, Snickare and Holter (ibid.) suggest that the idea of academia as (already) gender-equal is likely to be stronger in the Nordic countries than in, for instance, the US, reflecting the impression of the national value and identity tied to gender equality as a national branding (Larsen et al., 2021). This makes it a short leap to a postfeminist assumption (Budgeon, 2015) that *gender equality is already in place* and therefore *no further measures are needed* (Snickare & Holter, 2021, p. 40).

This way the widely accepted gender equality norm itself can be interpreted as already having solved the situation (Corneliussen & Seddighi, 2020a), making the existing gender inequality patterns disappear from view (Kaiser et al., 2013). It also makes gendered norms such as the male norm of IT appear as neutral for those who do not disturb this norm (Ahmed, 2012; Puwar, 2004). This increases the risk of the gender equality norm and policies becoming non-performatives (Ahmed, 2012), "meaning that the changes these measures and policies are meant to bring about are assumed to have been effected by the very fact of having a policy. In other words: nothing is done because a policy is in place" (Griffin & Vehviläinen, 2021, p. 7). This effect is recognized in the educational system, in working life, and in academia (Corneliussen, 2021a) and Griffin and Vehviläinen identifies this as one of the reasons that gender inequalities persist in the Nordic countries (2021).

The research revisited here illustrates that, despite the growing body of knowledge about barriers challenging girls' and women's participation in IT and other STEM fields, how to turn this into efficient and lasting initiatives to increase women's participation still appears as a mystery. The next chapters of this book will suggest another approach to these challenges by looking closer at how women who have already signed up for an IT education describe the decisive factors behind their own choice to enter masculine spaces of IT. The research literature documents that the challenge of increasing girls and young women's participation in IT is complex, with a spectre of barriers as well as factors affecting how girls and women find their way to a specialization in IT. Exploring women's narratives of their chronological pathway to IT, as we will do in the next chapter, opens for a wider understanding of which factors support women's journey from childhood to a university degree in IT. Petray et al. have criticized the oft-used metaphor of the leaky pipeline describing women leaving various stages of IT education and work, suggesting that part of the problem is that the pipeline metaphor "implies a singular pathway into engineering" (2019, p. 11). The women we will meet here illustrate not only that the conventional pathway to IT can be challenging for women to pursue, but also how they find alternative pathways that allows them to develop a sense of belonging in IT.

CHAPTER 3

Women's Chronological Pathways to IT Education

Abstract What are the key factors and driving forces that make women enter the fields of information technology (IT), despite the many gendered barriers revisited in Chap. 2? This chapter analyses the narratives of 24 women's chronological pathways from childhood to entering a university degree in IT. The chapter illustrates six different pathways that led the women to pursue a degree in IT, each analysed in terms of the positive drivers, including interest in IT, recruitment measures, an accidental choice, finding a safe platform in other disciplines, and a detour before discovering IT. Only one pathway identified the image of IT as suitable for women as a driving force. This, however, was shared by women from other countries, highlighting the specific cultural construction of the Norwegian women's narratives.

Keywords Chronological narratives • Recruiting women to IT • Women's study motivation IT • Career in technology • Women's alternative routes • Penalty round

INTRODUCTION: LOOKING FOR POSITIVE DRIVERS AND TURNING POINTS

What makes women enter information technology (IT), a field where there is still a notable gender gap across the western world (Eurostat, 2021a)? Chapter 2 revisited research documenting a wide set of gendered

© The Author(s) 2024
H. G. Corneliussen, *Reconstructions of Gender and Information Technology*, https://doi.org/10.1007/978-981-99-5187-1_3

barriers which challenged women's entries into IT (Master et al., 2016; Master & Meltzoff, 2020; Yates & Plagnol, 2022). These included gender structures and stereotypes picturing IT as a domain where men excel, making women question their abilities (Frieze & Quesenberry, 2019; Lewis et al., 2016; Margolis & Fisher, 2002). While boys' gateway through gaming is assumed to give an advantage also at university (Sevin & Decamp, 2016), dominating narratives question women's interest and belonging in IT (Faulkner, 2009; Sørensen, 2011) and women in male-dominated IT struggle to be recognized as professional (Faulkner, 2009; Watts, 2009). Research into study motivation has emphasized the close connection between students' expectations of mastering a subject, self-efficacy, and the value they ascribe to the field (González-Pérez et al., 2020; Sáinz & Eccles, 2012). The cultural association of IT with boys and men has been found to challenge both girls' and young women's ability belief and self-efficacy in this field (Barbieri et al., 2020; Chavatzia, 2017; Rohatgi et al., 2016). This effect is reinforced by women's lack of familiarity with IT, which gives them a poor foundation for assessing their own future performance in IT-related disciplines compared to men (Czopp et al., 2015; Spieler et al., 2019). Despite these and other barriers, women do find their way to IT, some through less conventional paths (Corneliussen & Seddighi, 2022; Lyon & Green, 2020). This chapter explores the pathways leading women to pursue a university education in IT while asking: what enabled them to get there, where did they find support, and what were the turning points that made them choose IT?

The chapter builds on in-depth interviews with and drawings made by 24 women who had set out on a career in technology (see the section on "Empirical Material" in Chap. 1). The analysis explores the narrative of the women's *chronological pathway from childhood until the decision of entering a university degree in IT*. The interviews involved explorative questions about the women's family background, experience with and interest in computers and IT since childhood, how they met IT in school and out of school, and experiences from IT at university. In this chapter, the analysis illustrates the different pathways that had led the women towards IT, guided by questions of where and when the women had become familiar with and interested in IT, and what made them start thinking about IT as a relevant career path. The interview guide also included more direct questions about gender in relation to IT; however, this topic did not surface in the analysis of what had led the women to IT, and I will return to this topic in Chap. 4.

The analysis below is rooted in a gender perspective of identity as constructed in and through social relations and interactions in social contexts (West & Zimmerman, 1987), combined with feminist technology studies emphasizing that this also involves technology and gender criteria for being recognized as IT experts (Nentwich & Kelan, 2014; Trauth & Quesenberry, 2007; see Chap. 1). Although gender is not particularly visible in the women's descriptions of positive factors and turning points that made them enter fields of IT, the gender perspective supports the understanding of how concepts such as self-efficacy, ability belief and value identified with disciplines of IT (González-Pérez et al., 2020; Master & Meltzoff, 2020, cf. Chap. 1) are both challenging for and challenged by the women.

A Chronological Drawing of the Pathways to IT

To answer the questions about what the women identified as decisive for their decision to study IT, we start with the drawing of the chronological pathways towards IT. This drawing was made on a predefined chart with two axes of age and school level and stages on the route to studying IT at university, as shown in Fig. 3.1. The chart thus identified four theoretical stages, from being unfamiliar with IT to becoming familiar and interested in IT, and finally making the decision to apply for a university-level IT degree. The drawing was used as a map during the interview for asking questions about what had caused the line to take on a specific direction.

Figure 3.1 summarizes the women's chronological drawings into three main types, showing that none of the women recalled having any insight

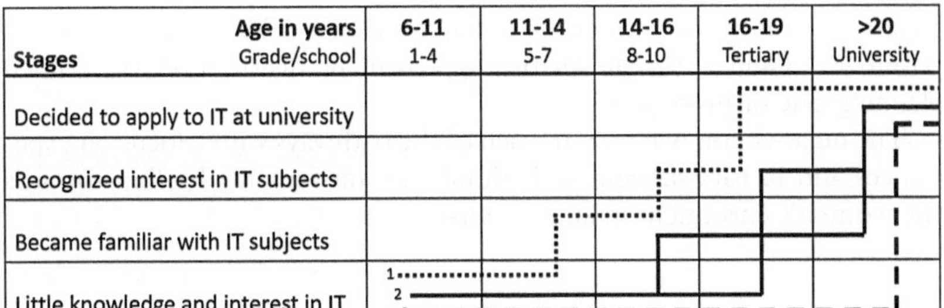

Stages	Age in years Grade/school	6-11 1-4	11-14 5-7	14-16 8-10	16-19 Tertiary	>20 University
Decided to apply to IT at university						
Recognized interest in IT subjects						
Became familiar with IT subjects						
Little knowledge and interest in IT						

Fig. 3.1 Illustration of the drawings that the 24 women made of their chronological pathways to IT

or interest in IT during the early years of primary school. From that point onwards, the women's narratives split into three main lines. Some of the women had developed an interest in IT during lower secondary school (line 1), while the majority developed insight and interest in IT gradually during the later stages (line 2). Some of the women, however, challenged the basic assumptions of the chart altogether; that the pathway to a university degree in IT involved certain levels of insight and interest. The third line represents a rather large group of women whose chronological pathways bypassed these stages while still ending up with a university degree in IT.

Hidden behind the lines in the figure there are many different experiences and variations that had led the women to study technology. In the next section, these variations will be elaborated and illustrated through six different pathways to IT education. These are a theoretical construction based on an analysis of the factors that led women to pursue careers in IT from around 2000 to 2020. They are not the only routes for women into fields of IT, and they can overlap, as illustrated by the narratives of six of the women who described experiences involving different pathways. Analysing the pathways will, however, help identify the factors that influenced the women's decisions to study IT, with each pathway illustrating certain factors that the women identified as vital for their decision. The first is characterized by an early interest in IT, while the second is shaped by recruitment efforts. The third pathway shows IT as an accidental choice, the fourth involves using an alternative discipline or interest as a stepping-stone into IT, and the fifth is characterized by women discovering IT as a relevant career path only after detouring through another career path. The final pathway involves women being encouraged to study IT *because* they were women; however, none of these grew up in Norway, thus highlighting the cultural construction of the Norwegian women's narratives. Figure 3.2 shows the approximate distribution of women claiming one or more pathways.

The next section will present each of the pathways with a focus on experiences and factors shaping and stimulating movements leading towards the women's enrolment in an IT course.

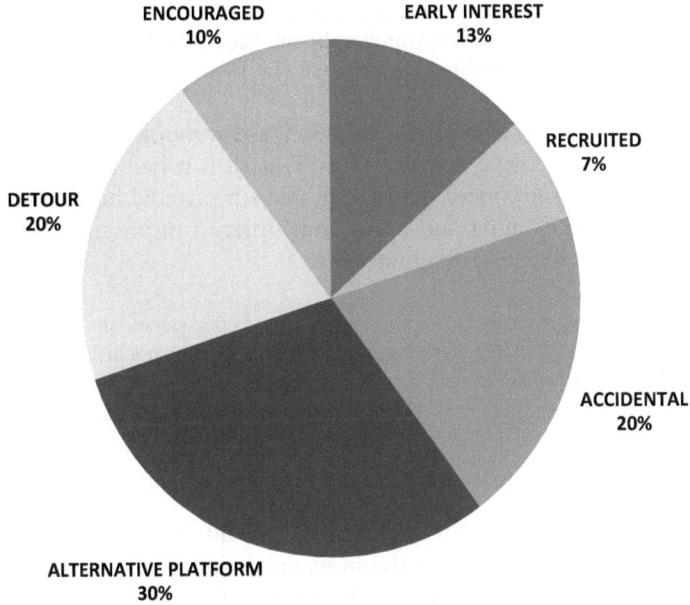

Fig. 3.2 The six pathways distributed around 30 stories of pursuing one of these routes. This includes all 24 women, six of whom claimed experiences involving two different pathways

PATHWAY 1: INTEREST IN IT AS A DRIVING FACTOR

The first pathway is shaped by women's early interest in IT, developed through encounters with technology. Reflecting a pattern of boys, on average, developing an interest in IT at an earlier stage than girls (Barker & Aspray, 2006), only four women recognized such an early interest. The typical scenario of the women's early encounters with technology involved computers, smartphones, or video games in their early teens, feeding a growing interest that gradually developed into a decision to study IT. Liv was one of those who followed such a path, and she recognized social activities, from online chat channels to gaming and programming, as the core of her interest in IT:

> My best friends were very interested in programming, and this affected me.
> I remember that when I was younger, before the smartphones, a lot of the
> social stuff happened through a computer. (Liv)

She had used early chat platforms as well as Facebook and doubted that
she would have become interested in IT at all if it had not started as a
social activity. A supportive network of peers motivated her engagement
with computers. Gaming and her friends' interest in programming triggered her own interest in programming:

> It started at upper secondary school, some friends had started making games
> on their own, and then I was inspired and wanted to learn a little too. Since
> I saw that they were doing it and I was playing video games myself, I was
> inspired to try it. That, in turn, made me want to take it to a professional
> level, and I searched online and found that this university had a degree in
> programming. (Liv)

A social environment of friends with similar interests played a vital role
in shaping her pathway, where the most important activity was programming. Before signing up for university, she had taught herself to program
in high-level programming languages such as Python and C++. It was difficult but fun, and it had given her an advantage when she started at
university:

> I felt that I was in front of the others, especially in the first semester, because
> there were many who did not know what [programming] was, really. It was
> great to have some prior knowledge and recognizing things, to avoid some
> of the pressure and the stress during the first months. (Liv)

Liv described her early encounters with IT through her social network
as being vital for her choice of IT, including certain similarities with images
of boys developing their IT skills through gaming and programming outside school. Thus, she seems to confirm the assumption that gaming can
support youth in developing knowledge about professional aspects of IT
(Sevin & Decamp, 2016). She also found that her early experience had
made her excel in the programming class, where all of her fellow students
were men.

The next three women also had an early interest in IT. Although only
one of them had learnt to program before university, all three defined their
interest in IT mainly through programming. They saw IT as creative and

a valuable skill for future work. Their narratives did not involve gaming, friends, or a supportive network, but described how they developed their interest in IT independently. One woman pointed to the increased importance of IT in society and its visibility through media as her main trigger:

> I thought it was cool. It seemed appealing based on what I had seen about IT on films and through entertainment. I knew it was difficult and demanding [to program]. [...] If you are good at it, however, it is actually a valuable thing to know. And I would have wanted to learn it anyway, because I'm a bit sceptical to how everyone wants to have smart homes and smart cars. How will that end, I wonder, since it can be hacked? (Marte)

Like several of the other women, she wanted to learn how to control technology, and this made programming a top priority. Since technology had entered the private sphere, it should be everybody's responsibility to take control: "If you bring a computer into your home, you must know how to use it one hundred percent and not just two percent" (Marte). The creative side of IT triggered these three women's interest in IT, and one of them saw IT as an alternative to becoming a writer:

> I like to create things. I was very fond of writing my own texts earlier. But being an artist can be very exhausting because it is not economically secure. Programming, however, ticks all the boxes I made for myself. (Marte)

Seeing programming as a combination of writing and creativity made this a perfect choice. However, when the women searched for a future direction in education, they found little support or guidance. None of them had received adequate information about this at school, which made other sources, such as internet searches, important. Lacking support, two of the women who wanted to work with web design ended up entering the wrong study program. They had been eager to learn about IT at an early stage, but in the absence of any available IT classes at school, they had to take care of this on their own. One of them signed up for a private IT class:

> It was just something I decided to do, it was not offered by the school. It was very challenging, because there are no one in my near vicinity or family that had studied IT. It was not very easy to get any help. And during the exam, I remember misunderstanding the assignment and therefore I didn't get a very good grade either. (Kristin)

Without support it was difficult. Although she was disappointed, she still dreamt of working with IT in a creative way:

> The reason I started at computer engineering, was that I wanted to become a web designer [...] I misunderstood the name "network architecture" and thought that it would be architecture and design. [...] I lost interest in computer engineering when I found out that I could not become a web designer with that education. (Kristin)

Choosing computer engineering by mistake, her motivation dropped when she realized this would not lead to web design. Her interest in IT, however, remained strong. She completed the bachelor's degree before changing to another IT degree in which she had a greater interest, where she was about to finish a PhD at the time of the interview.

The other woman who was interested in web design had started coding in HTML and was making webpages at age 12–13. She was dedicated to studying web design, but, without any guidance, she ended up pursuing a degree in graphic design instead: "I thought that I would become a web designer, because I thought that 'graphic design' would be more directed towards that, but then I realized that it was something completely different" (Elise). She completed her first bachelor's degree before changing to computer engineering, by which time she had lost interest in web design:

> If you are a web designer then perhaps you will make a logo or something, but what you do is not very important, in a way. However, if you become a computer engineer, then you can do a lot of important things, and that is more motivating. It gives endless possibilities. (Elise)

The women's narratives illustrate that they each faced challenges in finding and pursuing IT education due to a lack of support and guidance. They had received little information about IT education and career options through school. "I wish we had had visitors from the university, to learn about the different study programs", Elise said. She wished that they had learnt more about the requirements for applying because "that could have been something to work towards. But I was not motivated to work towards that, because I did not even know that it was a possibility" (Elise). Only after learning about computer engineering had she recognized a positive value of the field.

While the four women started with an early interest in IT, their experiences in pursuing an IT education differed, particularly in terms of access to a supportive network. Most activities identified as important happened outside the ordinary school system, making the schools rather irrelevant. In the case of one of the women, a family member working in IT was important for personifying IT as a career choice. The other three women did not have the same level of network or support and they had been left to navigate the educational landscape on their own. Their story is a reminder that critical challenges to women's participation in IT might not only be manifested in negative barriers, but can also be hidden by the absence of support.

Pathway 2: Recruited into IT Education

The first pathway included interest in IT, but no support from the educational system. The second pathway illustrates a reverse pattern: it involved no particular interest in IT; however, recruitment initiatives had a crucial role for their decision to study IT. Only two women emphasized such initiatives, both of which took place during their final year at upper secondary.

Solveig had already made up her mind to study either economy or law at university. Visiting an "information day" inviting all new applicants to university, made her change her mind:

> I had not thought about IT as an option before we were about to apply for higher education. At that time, I already knew what I was going to study, but then I followed a friend who was going to study engineering, to the lectures about that. That was when I realized this was a possibility too. I had put engineering far down on my list when we applied for university. Then we went on holiday and there I talked with someone who worked with IT. After that I changed the order on my list, moving engineering to the top the day before the deadline for applying to university. (Solveig)

Although she had informatics as an elective at upper secondary and had enjoyed the class, she had still not considered a career in IT: "I thought it was fun at school, but it never occurred to me that you can in fact study IT" (Solveig). Thus, even with IT at school, her narrative was characterized by the absence of schools providing insights that had made IT appear as a relevant course of study for her. Following the information meeting about computer engineering—a meeting that she accidentally joined, IT changed status from being irrelevant to becoming a possibility. Several successive events then made her gradually change her mind, until at the very last minute she made the final decision to make computer engineering her preferred choice when applying for university.

This narrative explained the line of her chronological drawing, which started out flat before moving directly from no interest and knowledge to a simultaneous point of interest and applying to university when she was 19 years old (see Fig. 3.1). While school had not had any impact on her aspiration to study IT, the turning point was the meeting, since this made her aware of IT as a study and career. The accidental character of participating in the information meeting is not a trivial detail of the narrative. Since the meeting was targeting those who already were interested in IT education, it was less likely to capture the interest of potential students such as her, who had not considered it relevant before. The next stage is also of interest as it reflects how young women might require several sources to develop their decision to study IT, as illustrated by the evaluation of a national recruitment campaign (Corneliussen, Seddighi, Simonsen, et al., 2021): once Solveig had become aware of IT as a potential study choice, she continued to investigate on her own until she gradually developed her knowledge about IT, which led to her ultimate decision to study IT engineering.

The other woman had participated in a recruitment initiative targeting women, and which was limited to those who had science as a major at upper secondary school:

> We participated at the "Girls' Day", which was a conference that we were invited to attend because we belonged to the science class. We were invited to [the university] to have a tour around campus and listening to several talks by people who worked in IT as well as students. [...] Then I thought "Hmm, maybe IT development is something I would want to work with." And that is when I got interested and decided that I wanted to apply for IT. (Ingrid)

Before attending the conference, Ingrid had considered medicine or nursing. She had not chosen IT as an elective at upper secondary because it was not a required subject and thus did not represent a strategic choice for keeping the door open to, for instance, medicine. When Ingrid participated in the Girls' Day, she started seeing IT in a new way that made it become a rival to medicine:

At the Girls' Day, what caught my interest was that you can work with society, but also with technology at the same time; that you can develop systems, for instance for hospitals, schools, NAV [the Norwegian Labour and Welfare Administration]. That you can use your technology education for something that is useful for society. And I realized that in medicine, you will have a job where you will be close to people and there will be a lot of blood and gore, and you need to know a lot about the body. I realized that I am not very good at that. I have more to offer, I am very good at math, and very good at logical thinking. (Ingrid)

Ingrid's experience highlights two important factors that can influence women's interest in a career in technology. Firstly, the realization that working with technology is applicable to many fields, and secondly that once she became familiar with IT, it emerged as a strong alternative to traditional health care careers, a realization that was shared by many of the other women.

Ingrid illustrates the challenge of acquiring a comprehensive overview of potential study choices, leading many young women to trust the most "obvious" choice, she believes: "either you become an economist, or you become a nurse" (Ingrid). For IT to become a visible and relevant study option for her, it required an invitation to learn about and experience IT. Once IT appeared relevant, she also recognized how her current competences in math and logical thinking might make IT a better choice for her. As an additional support, she recognized how meeting a female role model in the field had made her think differently about the relationship between girls and technology:

She made me see technology in a completely new way, with all the possibilities in it. [...] She made us think like; "yes, it is possible to study technology; it is totally fine to combine it with being a girl and not really having an interest in it." Or rather, you do have an interest in it, but it is possible because it is not only technology that defines you. (Ingrid)

She was struggling with the concept of working with technology that was not only about technology, and of being and not being interested in technology, complicated by expectations to girls' relationship to technology. Cracking the code for establishing a relationship with technology as a girl happened when she realized that technology could be interpreted in a much wider context and could be combined with nearly anything. Ingrid claimed that missing this information earlier was the main reason that she had not chosen IT before.

There are several lessons to learn from the narratives outlining the pathway that includes recruitment initiatives. Making IT visible as a relevant study choice is vital for putting it on the women's wish list, and it is not obvious that this happens through their everyday school experience. This makes recruitment initiatives important (Corneliussen, Seddighi, Simonsen, et al., 2021). Furthermore, motivating women to study IT can be effective up until the very last minute before they apply for university. Finally, Ingrid illustrates that interest in technology is closely associated with an image of a masculine technology relation that makes interest alone a fragile motivator for young women to navigate by. She also illustrates the importance of female role models as someone representing a different gendered version of technology interest (González-Pérez et al., 2020; Lang et al., 2020; Stout et al., 2011).

Pathway 3: An Unintended and Accidental Study Choice

While one of the women above illustrated elements of accidentally being at the right place at the right time, pathway 3 is shaped entirely by IT as an *unintended* and *accidental* study choice. Six women describe their pathway to IT as having starting without them having any intention of studying IT. This is visualized by the third line in Fig. 3.1, bypassing insight and interest, moving straight up to enrolling at university. One of these women decided to apply for a university degree; however, she had not decided what to study:

> I sat down with a list of university degrees in front of me, and then I just let my finger slide down the list. It stopped at an IT degree. I showed up in the first class. I was told to bring my own laptop, though I had minimal knowledge of that and went to [a chain for electronics] and bought a laptop and showed up at university. The first thing the teacher said was "Today we will program in Visual Basic". I had absolutely no idea what programming or Visual Basic was, so I hurried to google it while sitting there in the lecture, and I didn't understand anything. After class I visited my brother-in-law, who helped me get started with Visual Basic. And then I was really hooked, after only one class. Therefore, the study choice was completely random. (Gro)

A study choice can hardly become much more accidental than letting the index finger land blindly on IT in a list of university degrees. However,

Gro was not alone in finding her way to IT by accident. When applying to university it is possible to select several degrees, placing the first priority at the top. Tove, one of the other women, had put an IT degree further down on her list and then forgot about it, until she was admitted to the IT degree:

> I knew nothing about the subject before I started. I did not even know that I had applied for it, I had just applied for many different degrees at [university]. I just wanted to get into that university. (Tove)

It was not interest in IT that made her put it on the list when applying to university but rather other priorities such as getting admission to a specific campus. This narrative was gradually uncovered during the interview:

> Interviewer: So, you are not really interested in IT?
> Tove: No!
> Interviewer: What made you apply? Why did you not apply for a different subject?
> Tove: I did. I did not have it as a top priority at all. I had put in computer technology further up, and then I put a lot of weird subjects on the list. Because I really wanted to study math and physics, but I'm glad now that I did not end up there, because that would have turned out badly. I really just wanted to go to that campus. And then I wanted to become a civil engineer. I had no idea, I just wanted to start at a degree and then see. And then I started at computer engineering, which I enjoyed and did quite well the first year. Then I thought that I can finish this degree and that's the plan now. I'm good at it, sort of, so it's okay. But I do not know if I'm going to do it for the rest of my life, because it's not what I enjoy the most.

She identified her background in sciences and mathematics as a good background for mastering the computer engineering classes; however, as distinct from the account of Gro above, who was "hooked" from the first class of programming, Tove was unsure that she wanted to continue with computing. For her, being good at it was not the same as having an interest in it or enjoying it.

Several other women describe their entry into IT as either unintended or accidental, but who, once having started studying it, found that they enjoyed it and decided to continue. This suggests that educational choices are not always driven by an interest in a specific discipline or occupation but can be influenced by other priorities or accidental events. Ending up in an IT degree due to accidental events can lead to different outcomes, and while most of these women were satisfied with their choice, one of them remained uncertain whether or not she would stay on.

PATHWAY 4: AN ALTERNATIVE PLATFORM

The next pathway into IT is shaped not by a focus on IT, but rather reflects an investment within another discipline or topic of interests. The alternative field worked as a platform for claiming and establishing familiarity and a sense of mastery and belonging when enrolling in IT. The narratives of the nine women defining this pathway involved little interest in and knowledge about IT. Few of the women had learnt about IT at school and not many identified with a leisurely interest in computers. When describing why they chose an IT degree, they explained and justified their choice mainly with reference to another discipline:

> Mathematics was probably the strongest subject I had ever since I was a child. Therefore, my idea was that I had something that I could feel confident in, simultaneously as I would be able to learn something new. (Tonje)

The alternative disciplines were within fields that the women had already mastered, most often because they had the subject at school. The alternative subject thus already represented a platform for their self-efficacy, most often within a discipline not equally associated with men as most fields of IT. Thus, with the alternative platform boosting their confidence, the women could approach IT as a new and unfamiliar field while simultaneously keeping a link to their academic strength.

For some of the women it was the combination of their safe platform-discipline and IT that had caught their interest. "Since I was looking both at biology studies and computer science, it was the mixture that seemed interesting", Berit explained. She had considered either biology or computer science before deciding on bioinformatics as a combination of the two. Other women raised the same argument. Science alone was not tempting; however, when science subjects could be combined with IT, it became more attractive:

> I was at the info day for chemistry, and then I was like, "no, this is a bit old-fashioned, maybe a bit boring". But then I found the degree in bioinformatics on the internet. These are two things I'm interested in, the bio part and

the chemical part, and you can connect it with some IT. This was towards the end of upper secondary, so I was thinking that, OK, maybe I should start studying IT, because everyone has smartphones, everyone uses apps. Today, everything is digitalized. (Gunn)

For Gunn, it was not her interest in technology, but rather her *doubt* about sciences that had made her look for something else. She shared this doubt with some of the other women, such as Ida, who had chosen science and mathematics at upper secondary, but she struggled with these disciplines: "I did not really want to study math, because I felt that it might be a little too hard. Not really biology either, so I just decided to study computer science" (Ida). Ida's choice however, was also influenced by her father: "If it hadn't been for my father, who works in IT, I wouldn't have thought about it as a potential choice" (Ida). A father working in IT had made it visible as a potential study and career choice.

The turning point for many of these women, as Gunn illustrates, was the moment they realized that they could combine their already established skills with the new field of IT. For Gunn, like several of the other women, their background in sciences and mathematics was what had made them look towards technology. Lacking a background in technology, these other disciplines worked as a platform where they already had some skills:

I didn't have technology as a subject at school. In upper secondary I had science and I liked it then, but I did not like it so much that I wanted to take a bachelor's degree in only math or chemistry. [...] Then I realized that technology was a bit like a middle ground where you could both be creative and use science. (Ellen)

The women identified a wide set of disciplines, not only sciences and mathematics, but also social sciences, humanities, and arts as a platform for entering IT. The women's narratives thus illustrate how IT can be perceived as a good fit together with many different disciplines, topics of interest, skills, and values.

Many of the women were motivated to study IT because they saw it as necessary in today's society and believed that digitalization was increasingly important. One woman had already a professional experience and had come back to university because she felt that she needed to understand computing to continue doing her work: "We need to understand technology to understand [her professional field]. For me this is about being prepared for the new digital society" (Maja). Another woman wanted to learn how to control technology and "how we decide to design limits for technology", because this "will determine what society will look like in the future, which values we promote, which rights people will have" (Kari). Several of the women argue in a similar vein, claiming that their decisions to study IT were not solely driven by an interest in technology, but rather by the transformations in a society where digitalization took on an increasingly important role.

The women's narratives reveal that an interest in technology is not always the main factor in deciding to study IT. Many of them had limited knowledge and interest in IT before enrolling in university, and rather used their background in other disciplines as a safe platform to enter the less familiar field of IT. For some it was the combination of IT with another discipline with which they were already familiar, while, for others, IT appeared as a modern and relevant choice given its importance for work and education, for individuals, in private homes, and in today's society. The wide spectre of alternative disciplines and competences in which IT was considered important opened for a similarly wide set of competences that could provide a sense of mastery and belonging that could be transferred to IT. However, within the same narratives, IT often appeared as a *fragile choice*. Since few of the women had any experience from IT at school, they had no way of fully judging whether they would master or enjoy IT at university. The alternative platform supported the issue of mastery; however, whether they would enjoy IT was still an open question when they started at university.

Pathway 5: A Detour before "Discovering" IT

The next pathway was shaped by the women making a *detour*, most often in a different degree at university, before they *discovered* IT as interesting and fascinating. Realizing this, they changed direction and started over with a degree in IT. Five women contribute to our understanding of this pathway. While these women had not imagined studying IT when they were at school, the key shaping factor for this pathway was the input that made them change direction: something they saw, heard, or experienced that involved IT and made them start thinking of IT as a relevant study choice. Failing to receive (or pick up) this input during their school days, the detour was instrumental. For some women, it appeared as a *penalty round*; penalizing them for being unaware of IT when they made their first study choice as teenagers. For others, the detour rather allowed them to mature and grow out of a girls' culture with an effect of making IT appear as a less relevant study choice.

We met one of them above, as one of the women who had identified an early interest in IT; however, aiming for web design she had made a wrong choice and ended up in a non-tech design degree. While she was a student here, she had realized that she could attend a preliminary course which would qualify her to study computer engineering:

> I finished the bachelor's degree [in graphic design] and then I found out that you could take a preliminary course to get into computer engineering. And when I found out what computer engineering was, I thought that would be the right thing for me. Although I took a small detour, I ended up here in the end. (Elise)

Computer engineering had not been on her career radar before, again suggesting that the insight into IT disciplines is vital for how youth consider relevant study choices. Computer engineering had turned out to be "the education of her dreams". However, the detour via another bachelor's degree had consequences: "I will not do the full five-year degree of computer science, mostly because it will not pay off, considering the size of my loan compared to salary and income" (Elise). Leaving web design behind, her sense of computer engineering leading to more important work had left her satisfied about her new choice of education.

Different from her experience, the other four women in this group did not identify any previous interest in studying IT. All four completed secondary and upper secondary education without picking up any kind of input that triggered their interest in studying IT. It was not until they met concepts of computing at a much later stage that their interest was triggered. One of the women had first enrolled in a different engineering degree, where she had an introductory class to programming:

> I remember in the beginning when we had programming, I did not understand the way of thinking, but because I was forced to take that subject, I understood more and more that way of thinking. In the end I had a lot of fun and it turned out to be my best subject. Then I decided that I did not want to continue with [the first degree] but instead wanted to start at computer science. (Sofie)

This unplanned meeting with programming, struggling at first, but then realizing that she was good at it, made her decide to change into IT. Her chronological drawing had not moved above the first stage of ignorance and lack of interest in IT, until it made an abrupt turn upwards, motivated by discovering her abilities and pleasure in programming. And she was not alone in being recruited through an unexpected meeting with programming, as the next woman shows:

> I cannot explain the joy I got from having an introduction to programming. [...] For me, it has been one of the coolest things. It's not like any other subject I've had. It was the absolute coolest thing about joining that study. (Anna)

Programming has often been described as one of the things putting women off IT education (Denning & McGettrick, 2005; Jethwani et al., 2016). Quite the opposite was true for these women: the fascination and pleasure of this experience played a key role in their narratives, giving a new value to computing and explaining the change of direction:

> We had the introductory subject to IT [...] and I thought it was the most fun subject. It was more fun than the subject of chemistry because I had never had IT before. Therefore, after a year I chose to switch to an IT degree. (Tonje)

The women's emphasis on programming as a completely new and unknown subject reflects just how little insight they had received through school. The detour was instrumental in establishing such an insight, representing an extra round of studying where they discovered IT. One of the women even suggested that this was typical for her fellow female students; that choosing IT was not something they had been prepared for at school:

> I think almost all the girls in the class chose this degree at random. And it's a bit of a shame that it's like that, because seeing that I found IT so much fun when I started here, I would probably have been hooked the same way when I went to upper secondary. And then I could have been saved for an extra year of study loan. (Lene)

For her the detour had worked as a penalty round; an extra round adding an extra cost to her career development. She could have been recruited earlier, she thinks, but nobody had invited her to explore IT at school: "At lower and upper secondary school we did not have any IT class. There was nothing, so it is not at all strange that I did not become interested in it" (Lene).

IT was not on these women's horizon when they were making plans for their future career. On the one side, the weak importance of school in these narratives—a characteristic that these women share with many of the other women—suggest that many women are not successfully recruited through activities at school. The women's change of direction, on the other side, documents that it is never too late to be recruited to IT. The various encounters with IT, in particular hands-on experiences with programming, had made the women change their perception about IT completely. The combination of little input at school with the much later unexpected encounters that brought new meaning to IT shapes this pathway involving a detour and late entry into IT.

PATHWAY 6: ENCOURAGED BECAUSE IT IS SUITABLE FOR GIRLS

There is something cultural about it. It was one of the most appropriate studies. Perhaps not the most appropriate, but one of the most appropriate studies for girls in [her country]. I knew this because of my sister. She encouraged me. I think she chose IT because she had a teacher who encouraged her.

Three women expressed similar experiences; IT was a subject they had been recommended and encouraged to study. The three women had grown up in countries further south and east at the fringe of Europe. They had come to Norway for study at master's and PhD level, thus, they saw themselves as visitors in Norway. Including them in the study gave an opportunity to see not only how cultural aspects are shaping the foreign women's pathways, but also how the Norwegian women's experiences were also culturally shaped.

A vital element of the foreign women's chronological narratives was the experience of being *encouraged* to study IT; by family, at school, and in general by a cultural discourse about IT as an appropriate study choice for girls and women in their home countries. Another reason for being recommended and encouraged to study IT in their experience, had been skills in mathematics, which they identified as their strongest subject at school:

Aisha: We have a lot of math in [my home country] and they suggest that if you are strong in mathematics, you can be a strong candidate for IT studies. Interviewer: Yes, so if you are good at math, you will be guided towards that subject? Aisha: Yes.

One of the women emphasized that IT studies attract the very best students in her home country. Being one of the best students in her class had therefore motivated her to choose IT:

> I was motivated because it is a trend and technology is everywhere. And what is happening with the universities now, is that IT departments are accepting students with the best results from [lower education], thus it is the best students who study IT. These are also the reasons why I chose IT. I had the best results at school, and that's how I chose direction. (Dafina)

The narratives from the three foreign women are very different from those of the Norwegian women. They reflected on this difference themselves, describing a huge surprise when they arrived in Norway and found the proportion of women in IT to be very low. Their surprise was not related to Norway as a country recognized as gender equal, but rather as a strange situation across Europe in general compared to the much better gender balance in their home countries:

> It was really strange for me when I came to Norway, since some fields of engineering in [country] are very male dominated, however, computer science was more 50–50 with men and women. I think it's a cultural thing. And I understand that it is the same across Europe; girls are not interested in computer science. [...] It was a shock to me: "Really, am I the only girl in this class?" (Aisha)

Coming to Norway for a university degree and finding themselves to be one of the few women in IT appeared different from their experience in their home country; however, it also affected them:

> Even the professors who come to teach the courses [are men]; every day you only see men. This means that you also give the impression to the students, that this field is dominated by men and therefore it is only for men. (Dafina)

Even those women who had grown up in a culture where at least certain fields of IT were thought of as gender-equal fields, were affected by mainly seeing men at the IT department at the Norwegian universities. They started to question their own participation, and their former feeling of belonging was threatened by a sense of being an outsider in what they experienced as a mainly male-populated IT department at university.

While the aim of this study was not to explore the co-construction of gender and technology in a *cross-cultural perspective*, the foreign participants challenged this by highlighting the cultural construction of the Norwegian women's narratives. Their experiences emphasize that the cultural construction of IT as a masculine field is a result neither of inherent qualities of technology nor of innate skills in men. They also illustrate the negative experience of only seeing men operating in a field and the immediate effect this had in making them start questioning their belonging in IT.

Looking back at the Norwegian women's narratives with lenses sharpened by the foreign women's stories highlights the cultural aspects of Norwegian women's experiences, and in particular that none of them had been encouraged to think of IT *because they were women*. The foreign women thus put the Norwegian women's narrative into a perspective that suggests that it could have been different.

LESSONS FROM THE PATHWAYS

The main goal of the analysis above was to explore the factors that the women identified as vital in shaping their pathways to IT, from childhood to a university degree in IT. The six pathways illustrate that background and motivation for studying IT can differ considerably from woman to woman, reminding us that women are not a single homogenous group (Trauth & Quesenberry, 2007). While few women had identified interest in IT in their teenage years, many of their narratives showed that interest in IT had not been necessary for their choice of studying IT. Instead, interest in a wide spectre of topics and disciplines, or an interest in society's transformation due to digitalization, had led to an interest in studying IT. These other fields and disciplines had also supported their self-efficacy and trust in mastering IT, illustrating that not only IT or other STEM disciplines support ability belief in IT. The women's narratives thus also challenge some of the main assumptions of theories emphasizing the importance of self-efficacy and interest in the discipline in question (Eccles, 2009; Master & Meltzoff, 2020; Rohatgi et al., 2016).

One common factor across the pathways was that the women needed some kind of input, insight, or experience that made IT seem relevant. Having a family member presence in the industry made IT visible as a potential choice for some of the women, while only one woman identified friends as having played an important role. Only two women had experienced that recruitment initiatives had made them want to study IT, while the more common experience for most of the women across the pathways was of school as being either irrelevant or making IT seem less tempting as a career path, similar to findings in other studies (Alshahrani et al., 2018; Engström, 2018). The women illustrated both the benefit of having multiple sources for inspiration, and the abrupt change from a more gender-traditional study choice to IT once the possibility had been presented for them, for instance in recruitment initiatives (cf. Chap. 2, Corneliussen, Seddighi, Simonsen, et al., 2021). Thus, while the women's narratives emphasize that they needed input to make IT appear as a relevant study choice, their stories also point to scarcity in such support. The foreign women's narrative of being invited to study IT because it was considered suitable for women puts the Norwegian women's experiences, in particular their lack of support and encouragement, into sharp relief.

While the six pathways analysed here are neither the only routes, and nor are they exclusive to women, the women's narratives reflect how they had experienced this as a journey through a gendered landscape where being a woman had certain implications. The stories about how gender affected their experience had, however, not surfaced when analysing the factors that had *enabled* women to enter fields of IT. Though the sample here is limited, it is still disheartening that merely a handful of the women had chosen IT because they had an early interest in it or recognized being encouraged or recruited. This will be further explored in the next chapter, which will engage more directly with the questions of how the pathways can be understood in relation to the cultural construction of IT as a gendered field.

CHAPTER 4

Girl Power: Reconstructing the Gendered Space of IT

Abstract This chapter explores how women navigate and challenge gendered stereotypes defining IT as a masculine space. Most of the women had approached IT with limited insights. This made gender stereotypes, including a male-dominated storyline of gamers, geeks, and hackers, central to their early perceptions of the field. However, once they learnt more about IT, they started defining their own strengths and belonging in the field. The women's experiences are analysed in light of Puwar's metaphor of "space invaders", highlighting how women appear as "bodies out of place" in a masculine space of IT. The space invader identity is also productive, and the women reconfigure the notion of IT as a wider and more open space where also women can be considered insiders.

Keywords Female role models • Male storyline in IT • Reconstructions of IT • Space invaders • Women's interest in IT • Women's visibility in IT

Introduction: Space Invader Experiences

The previous chapter explored what had enabled women to enter university-level IT programs. When focusing on positive drivers and turning points that enabled and contributed to women's arrival at the start of a career in IT, gender remained mainly unspoken. Yet, gender was inextricably entangled in the women's experiences. Few others than the foreign

© The Author(s) 2024
H. G. Corneliussen, *Reconstructions of Gender and Information Technology*, https://doi.org/10.1007/978-981-99-5187-1_4

women recounted experiences where being a girl had led them onto a pathway to IT. In the Norwegian women's narratives, gender rather came up when they described their uncertainty about studying IT, with questions such as: *if you only see men, will I fit?* This chapter takes a closer look at how IT was perceived as a gendered space wherein being a woman made a difference to and shaped their experiences. The questions pursued here are: how do women navigate the gendered landscape of IT? How were they challenged by and how did they themselves challenge the gender structures and stereotypes of IT?

The analysis of the women's navigation of the gendered spaces of IT is inspired by Puwar's concept of *space invader* as a metaphor for women and racial minorities entering historically male and white spaces that had previously been less available to them (2004). While male bodies appear to represent neutral positions, the space invader metaphor highlights how such reserved spaces are shaped "through what has been constructed out" (Puwar, 2004, p. 1). The paradox of the space invader appears when "bodies out of place" enter these reserved spaces; being there, enjoying it, but not fully belonging there (ibid.). Challenged by the culturally constructed norm, the space invader often meets doubt about her belonging and "super-surveillance" questioning her skills (Puwar, 2004, p. 11).

Here I use this concept to explore the women's experiences when entering higher education in IT in disciplines within technology and sciences departments, that is, spaces of IT inhabited mostly by men in Norway (The Norwegian Universities and Colleges Admission Service, 2022). Many of the women shared the feeling of disturbing a masculine norm of IT, of not fitting the main narrative and the expectations to inhabitants of such spaces. The women's experiences thus reflect research illustrating how being a woman and interested in IT appear as a contradiction in most western cultures (Chow & Charles, 2019; Nentwich & Kelan, 2014). Gender stereotypes not only challenge young women's association with and self-efficacy in IT (Borgonovi et al., 2018; Chavatzia, 2017; Frieze & Quesenberry, 2019; Watts, 2009) but also their ability to identify female role models (Arnold et al., 2021; Corneliussen et al., 2019). The women are, however, not passive bystanders. By disturbing the norm, they contribute to identifying the social construction of the male spaces of IT. Simultaneously, they challenge the masculine norm by suggesting new ways of understanding IT, what it represents, and who can pass as expert in the field.

Below I will start by exploring how the women's understanding of IT changed as they developed their knowledge about the field and how this affected their perception of the gendering of this space. The women also negotiated the gendered norm, finding ways for bridging the seemingly contradictory positions of *being a girl* and being involved in IT.

GAMERS, GEEKS, AND HACKERS—IDENTIFYING THE NORM AS MASCULINE

Few of the women we met in the previous chapter had recognized IT as a study or career option at lower or upper secondary school, and some not even when starting at university. One reason for this was that their initial understanding reflected narratives recognized in previous research, of IT as a world mainly populated by men and little inviting for women (Faulkner, 2009; Lewis et al., 2016; Sørensen, 2011). The women's perceptions developed over time and with experience, and their later understanding supported a revised vision of who belongs in spaces of IT. When the women developed their sense of belonging in this field, it was entangled with their reconstruction of IT as a more open and inclusive space. However, their initial understanding provides important insights into how the women negotiated their belonging in IT as space invaders disturbing what they perceived as a masculine norm of IT, thus highlighting some of the main features of the historical and conceptual constructions of IT as a gendered field (Puwar, 2004).

The women's different pathways to IT analysed in the previous chapter illustrated that most of them had limited experience with IT and therefore recognized their lack of knowledge about IT as a discipline and occupation when they entered university. Most of the women therefore had to navigate the educational landscape based on their initial understanding that included stereotypes defining IT as a masculine field: "*It has become more like a boy's subject, and it seems like only guys do computing and stuff like that*" (Berit). The male characters that women expected to find in IT programs were labelled "gamers", "geeks", or "hackers", and they were assumed to have an intense but also a narrow interest in video games and programming, as Elise illustrates: "*I really expected [the study environment] to be a little more monotonous with the typical nerd and hoodie and the 'gaming all day long' type*". This combination of nerd, hoodie, and gamer that dominated women's initial understanding of IT had made it difficult for them to imagine themselves in IT: "*I associate it with people who like to game a lot, who sit like that in their room [...] So I thought it wasn't for me at all*" (Gunn).

This gendered pre-understanding in the women's narratives can be described as a *storyline* following a certain pattern (Søndergaard, 2002): boys play games; gamers need to learn to program; boys develop programming knowledge before entering higher IT education. Girls and women do not fit the image of young men in hoodies. This initial image, which dominated the women's perception of IT, did not leave much space for the young women to imagine their own belonging in IT. Furthermore, lacking the background associated with the male gamer was perceived as an obstacle by most of the women, since they could not recognize themselves as correctly equipped to enter IT: "*I had never programmed before in my life*" (Ingrid). While trying something new might be scary, doing so along fellow students that the women assumed were already skilled programmers added strain to the situation. Thus, the women did not talk much about interest in IT as shaping their decision to study IT. Instead, they pointed to the double challenge they experienced, of neither fitting the image of the male insiders in spaces of IT, nor holding the skills or competence they associated with this position.

CLAIMING VISIBILITY—CLAIMING SPACE

Gradually, as the women came into closer contact with IT and the relevant environments, they developed their perception in ways that redefined IT as a place where also women could belong. The process by which the women started to define their own belonging was targeting their initial perception, including the male-dominated storyline, in several ways; by claiming women's visibility in spaces of IT, and emphasizing alternative interests, strengths, and competences as important. While the women challenged the hegemony of the male storyline by participating in spaces of IT, their narratives illustrate that this was not a simple or straightforward process, but rather one including a risk of backlashes where a male norm was reinstated. For some of them, such as Tove, the first meeting with university still seemed to confirm their initial feeling of not belonging there: "*When I showed up the first day, I didn't feel like I fit in [...]. There were a lot of typical gamers there, and I've never played videogames in my entire life, right. So I didn't really feel at home at that point*" (Tove).

For most of the women this image gradually changed. A first step was to identify other types of people than the gamers among their fellow students. Several of the women pointed out, with a certain surprise, that they had found *"normal people that you can talk with"* (Berit) and *"people like me"* (Tonje) among their fellow students:

> When I started, I was a bit like "no, these are computer people, they are not the ones I hang out with, I am not a gamer". Because I imagined a bunch of gamers, but it was rather like "I can actually fit in here". People are so different here. So yes, it was completely different from what I expected. (Gunn)

While the hooded gamer still played a part in the women's perception of IT, also other types of people came into view. For most of them, a presence and visibility of other women was important for them to feel welcome: *"When I started studying, it was very important for me to see other women. When I saw them there, I felt that I could be there too"* (Sofie). Finding other women in spaces of IT enabled the women to start developing their own sense of belonging. Several of the women had initially expected and feared that they would be the only woman in an all-male space of IT. Seeing other women in IT made a big difference, because *"you do not have to feel like you're a complete outsider even if you're a girl: you feel that you can be part of a group of girls"*, Gunn said. Several of the universities had acted as a facilitator for developing a community of women:

> The first day we had a girls' day, where all the girls got to meet and to know each other [...]. From day one I felt like I was almost going to class with only girls. Because they are the ones I mostly talk to and sit together with. I can't say I think much about it really, that there are mostly boys here. (Ingrid)

Whether formally organized or reflecting informal practices, most of the women appreciated and participated in communities of women when they started studying IT. The high visibility of women contributed to normalizing women's presence while also reducing the visibility of the male dominance: *"We don't notice [that there are few women], since the girls are the ones that you see in the reading room, so it doesn't feel like there are fewer girls than boys here"* (Berit). While many of the women appreciated the surprise of finding a rather large group of female students, the same was not the case for the staff at the IT departments. This reflected a generally low percentage, with less than 20% female professors in technology disciplines at Norwegian universities in 2020 (Steine et al., 2020). Noticing this, the women expressed disappointment about the lack of women among professors and lecturers as well as among student assistants and study group leaders. Anna had been "*upset about how few female group leaders and seminar leaders there are*", while Tove was proud to become a student assistant "*because a girl has never been a student assistant before.*"

Different from Kanter's notion of "token" pointing to professional women's experience of being devalued because they are seen as representing women, rather than a profession (Kanter, [1977] 1993), several of the women used their visibility in a more assertive way, as an advantage: "*If I am the only woman in a conference hall, then everyone knows who I am, even though I do not know who they are. If I say something then, I will be heard*" (Camilla). Putting themselves in the limelight in male-dominated IT contexts became a tool for making women visible:

> I'm one of those who raise their hands in class [...] I make up nearly 20 percent of all the girls in the hall, whereas you never see 20 percent of the boys speaking in class. [...] So I take on that task; I raise my hand, even though it may be a little unnecessary at times. (Anna)

The visibility of women in IT reading rooms and lecture halls, as lecturers and role models, contributed to a collective feeling of empowerment among the women, justifying their presence and participation in spaces of IT. Acts such as raising their hand or their voice in male-dominated spaces of IT had become a conscious way of flagging that women were participating, both as women and as professionals.

Women's visibility was vital for the women to consider themselves as qualified for studying and working with IT:

> It was very important for me to see that others with whom I could identify, people similar to me, achieved things here. Because if you only had the stereotype of these men, then I don't think I had felt that this was suitable for me, because my subconsciousness would have been like "no, perhaps you don't fit here". (Gunn)

Seeing other women with a career in IT communicated a message about the possibility for women to succeed: "*you feel that if they can make it, then maybe I can make it too*" (Gunn). Many of the women emphasised the importance that not only men, but also women, were available as role models for technology professions: "*if you only see men [...] it is not so easy to imagine 'being him' some years from now*" (Ingeborg). Female role models in male-dominated spaces represent counter-stereotypical images that can work as door openers for young women who find it easier to associate themselves with other women rather than with men (González-Pérez et al., 2020). Female role models are important because girls "*have a very different pathway—it is very different to be a girl*" facing the male-dominated world of technology, Ingeborg explained.

Cracking the Code of Being a Girl in IT

The women's experiences as "bodies out of place" in the masculine spaces of IT had made women's visibility important, and female role models were also vital for cracking the code of being a girl in IT. In Chap. 3 we saw how Ingrid had been struggling with bridging the positions of *being interested in IT* and *being a girl*, which for her had appeared as mutually exclusive. A woman featuring as a female role model in a recruitment event targeting young women, had helped Ingrid to start thinking differently about how to make these positions meet. She was still wavering between "*being a girl and not really having an interest in it*" and of "*having an interest*" that was made possible by the thought that it was "*not only technology that defines you*" (Ingrid). Gender and interest in IT are closely knitted together in the gendered norm of IT, illustrating the challenge for girls and women entering spaces of IT when the masculine norm seems to disqualify them simply for being female. For Ingrid, seeing a female role model who had not only succeeded in a tech career, but also expressed her interest in technology, was instrumental for her vision of the possibility of bridging these seemingly contradictory positions. This example also illustrates the potential effect that female role models can have for girls who had never fully realized that the combination of girls and technology was an option.

The importance of female role models was confirmed by most of the women. Female role models in technology have, however, not been readily available for many of the women: "*I haven't really had a role model, such a female role model, because it hasn't existed*" (Camilla). Some women resort to "substitute" role models from other fields and areas of life when they cannot identify women as role models in technology (Corneliussen et al., 2019). Others rather found female role models on social media, such as the following illustrates:

> After I started at the programming degree, I have sought out many role models online. On Instagram, for example, there have been several female pro-

grammers [...] When seeing that they are doing so well, that has been a major inspiration for me to continue [...] So whenever I've felt insecure about my choice, then I've kind of looked at them and felt a little more confident. (Liv)

Even for the women who could identify with a stereotypical male trajectory to IT, growing up with gaming and programming, it was important to identify women as *someone like her* to support their feeling of belonging in IT.

Although most of the women appreciated the visibility of women and female role models, their narratives indicate that women had not been particularly visible outside the arenas of IT education. As we saw in Chap. 3, only two women emphasized recruitment activities involving female role models, and none of the women referred to meetings with other organizations or networks for women and technology. This relative invisibility of women in tech is one of the mechanisms that makes it necessary for each new generation of young women to establish a relationship between being a girl and a career in IT.

Alternative Interests, Motivations, and Competences

While the women's experiences paved the way for a wider notion of participation in IT, the image of the hooded gamer still appeared important in the women's perception of IT. Gaming leading to programming was still assumed to be important and formative for the level of skills among students. Reinforced by a widespread perception of programming as a core field in IT, this remained an insurmountable challenge for those women who did not have a similar background, because they could not see themselves competing with men in this field: "*There are two or three boys in my class who had a lot of experience from before. They are not the majority, but they are so skilled that it feels completely unattainable to be as good as them*" (Lene). Paradoxically, it was programming that made many of the women fall in love with IT. Anna, for example, illustrated this when saying, "*I can't explain the joy I got from my first programming class*". Many of the women expressed pleasure, fascination, and spoke of becoming "hooked" on programming. This was followed by an expression of their sadness for not having learnt it before, as illustrated by Gro: "*If I had known about the possibilities before, I would have sat down and started programming right away*".

The women's limited background with IT was also reflected in how they considered themselves *suitable* for working with IT. None of them identified themselves as being good at IT and few described programming as one of their strengths. Yet the women's self-assessment responded to the key features of the male storyline, which involved interest, motivation, and competence:

I think I fit here because I like learning new things all the time. Without really knowing when I was a kid, I was into everything that was technological and, sort of, doing a little bit myself. [...] So, I feel I fit in because I'm motivated [...]. Not because I'm the best at programming or the best in math, but I just feel that I have the motivation that is needed. (Gunn)

Gunn's emphasis on being motivated, and her childhood interest in technology, resembles many of the other women's descriptions of their interest and motivation—exactly the things that the male storyline ascribes to men rather than to women. However, their interest is not primarily in IT—some even refused having an interest in IT at all, and rather defined it as a more general interest in society or, as Ellen, in the world: "*I am interested in the world, and technology is a large part of the world*". The women translated their motivation to learn about IT to a wish of being prepared for the "*digital society*" (Maja), to participate in decisions with the aim to contain and limit the effects of digital technology, and to be able to solve societal challenges. The value that women identified with IT was thus not limited to technology but rather included a wide range of goals and values, from supporting individuals' ability to deal with cyber threats to making them able to participate in the ongoing digitalization of previously non-tech disciplines and occupations (Corneliussen & Seddighi, 2022).

A similar picture emerges when they explain their professional strengths; very few of the women identified any background in IT, while most of them identified their professional strengths in sciences and mathematics, or in social sciences or business. For many of the women, this had produced a safe platform for entering IT, as we saw in Chap. 3 when Tonje explained that mathematics made her confident "*while learning something new*", referring to IT. The women described a wide set of skills they identified as relevant for IT such as language, arts, or a combination of competences with nearly any field such as *the world*:

I have always been very good at science. I kind of felt it was the same type of thinking. And I am very good at languages. [...] I think in a way that programming is just learning a new language. [...]. So I really like the combination. In a way it's like writing a text that you have to write in the best way possible. At the same time, there is a very hard logic to it. (Ingrid)

The women emphasized that IT can be relevant nearly anywhere, whether you work in a store or in a hospital, many using recognizable female-dominated workplaces as examples.

The women's descriptions of their strengths can be interpreted as a response to the male storyline as it confirms that also they had interests, motivations, and competences necessary in IT, albeit ones that were different from those associated with the figure of the hooded gamer. The safe platform made it possible for the women to establish self-efficacy by recognizing that their own strengths and competences could also be described as relevant for IT. Furthermore, the *alternative* platform simultaneously allowed the women to identify their strength in IT without entering into direct competition with the (image of) male programmers. Although many of the women expressed a love and deep fascination for programming, this still remained a male domain where the women doubted that they could compete with the men who "*have been programming since they were young*" (Elise). By suggesting a wider understanding of what and who were suitable for IT, the women avoided competing with the male storyline. The wider perception of IT reflected the women's strengths, rather than (only) the image of the male hooded gamer. This new understanding of IT also challenged one of the assumptions that was entangled in their initial understanding, in which knowledge about IT, and in particular programming, appeared as a prerequisite for entering IT education, similar to perceptions identified also in other western countries (Margolis & Fisher, 2002; Yates & Plagnol, 2022). The women's experiences from university had taught them that it was possible to start from scratch, and some of the women even develop this into a form of *discursive protection* against stereotypical assumptions about who fits in IT:

> Programming is not like any subject I ever had before. [...] It was like a whole new thing. And you had all the prerequisites to succeed because it was just completely new. It's not a subject you can say "no, I'm not going to be able to do this because I'm not good at math". Or "no, I'm not going to be able to do this because I'm not good at English". It's not like you can say that because it doesn't resemble any of those things. So programming is really fun. (Anna)

Seeing skills in IT as a prerequisite for studying IT had challenged the women's entry into IT because it appeared to identify their outsider position as one that was impossible to change. However, as Anna illustrates, she not only refused to be judged by her earlier (lack of) experience, but also pointed out that her lack of a pre-study competence reflected that schools had not offered insights into programming. Other women used similar arguments to explain that their lack of *early* interest in IT was a result of not having access to knowledge about IT or the option to developing skills in programming, because this had not been on offer at school: "*At secondary school, we did not even have an IT class. There was nothing. So, it's no wonder that I didn't become interested in IT when there was nothing available*" (Lene). Several of the women used this technique of identifying IT as completely new, referring to Astrid Lindgren's children's book character Pippi Longstocking and her take on life to illustrate their situation: "*I have never tried that before, so I think I should definitely be able to do that*".

The women's descriptions of their alternative strengths and motivations can be seen as a response to the male norm in IT. While most of the women did not challenge the core of the masculine relationship with IT, they rather introduced a reconstructed notion of IT that challenged the assumption that only individuals with a specific background, interests, or gender can succeed in IT. Pippi, the "strongest girl in the world" who would fearlessly take on any task she had never tried before, worked as a motivation and encouragement to take on a new task with an optimistic encouragement for their self-efficacy in a new field.

The Space Invaders' Penalty Round

Most of the women had initially felt that they did not fully belong, not due to formal barriers, but because of a cultural association between men and IT, which defined it as a masculine space that made the women question their background skills and whether or not they would fit in. This was entangled with the pathways discussed in Chap. 3, of basing their belonging in an alternative field, making the choice accidentally, detours and late arrivals; strategies that can be understood in light of the space invader metaphor. Leaning on a less male-dominated discipline such as social science, humanities, arts, or even mathematics made women's participation in IT less "gender inauthentic" (Faulkner, 2009). Describing their entry point as random took the edge off the choice, as if they were not competing head-on with the male norm. Some of the women who had not been recruited to or identified IT during their school days and had first started on a less male-dominated education before "discovering" IT, described this almost as a *penalty round* shaped by the lack of any invitation for them to learn about IT. The women's experiences, and their unconventional pathways into IT, reflect the gendered structures and stereotypes that specified IT as a male space and a less relevant career choice for young women. While pursuing unconventional pathways to IT is not reserved to women alone, the women's chronological narratives highlight how their movements from childhood to entering IT at university had been affected and motivated by IT as being inextricably linked to men, gaming, and programming.

The hesitation in many of the women's entry into spaces of IT involved a double uncertainty; they were unsure whether they would fit among the already skilled male students, but they were also unsure as to whether or not they would *enjoy* being there, referring to a completely new type of competence that they had not met at school. Arriving as outsiders, and not seeing themselves as fully fit with the available (male) position, made many of the women describe their entry with a certain reservation: "*Since I have no experience I do not know whether I will like this or not, and therefore I'm just giving it a try*" (Astrid). The women's uncertainty reflects that they had been lacking insight that could have told them whether they would like to study IT; thus many claim to be "just testing", appearing as visitors, shopping for new experiences, albeit without investing too much. This rhetoric takes the sting off the intrusion of the space invader's presence in the space where they do not fully belong. Furthermore, it suggests that many of the women had at first not imagined themselves in an insider position, leading to an uncertainty that was often expressed together with a readiness to move on to something else:

> I want to be completely honest and say that I was not very interested before I started [...] So it was like, OK, if I absolutely do not like this, then I can always switch to something else. (Gunn)

Being prepared to backtrack their choice worked as a safety net. Some of the women were still not completely sure that IT was a good choice for them. "*I probably do not have a very high interest in the subject*", Kari said, as she had started to question her own interest in IT, echoing Margolis and Fisher's 20-year-old study from the US (2002).

Reviewing the women's chronological narratives of how they had come to enter fields of IT through the concept of the space invader, illustrates how the pattern of unconventional pathways had been shaped by gendered experiences. Their lack of knowledge and insight in IT, combined with gendered stereotypes and structures that made IT seem a less welcoming career choice for young women, explains their hesitation and their doubt about choosing IT.

Is it Enough to Feel like an Insider?

Most of the women made the initial choice to study IT while they were still seeing IT mainly through a stereotypical perception, including the male storyline of hooded gamers. Furthermore, most of the women developed interest and confidence in their abilities once they entered university and were able to identify their own interests, motivations, and competences as relevant and important for IT, despite being different from the male-centric storyline. They also found support and community in other women, which helped them feel less outnumbered by male students. Their active participation in social and professional arenas to increase visibility and create positive images of women in IT, supported the empowerment of themselves and other women in the field. The women thus gradually changed their notion of IT to a more open and inclusive space that accepted alternative positions where women also could pass as insiders. Some of the women had, however, from the very start approached the IT study with a perception of themselves as suitable insiders. One of these was Liv, whom we met in Chap. 3 and who had a background with similarities to boys' leisurely and playful engagement with IT that included games and programming. This had made her a more skilled programmer than her fellow students on her bachelor's degree, all of whom were men. While this made her identify with an insider position in IT, she still found that her belonging was questioned:

> There have been situations where I feel that people are looking down at me because I am a girl. They have the stereotype about girls not knowing anything about computing. I often feel that I must prove myself for them and show that I am skilled and know my stuff. (Liv)

Her competence in IT had not secured her position as an insider because she did not conform with the male norm. Being recognized as a girl endangered her sense of belonging in IT, and she felt that she continuously had to renegotiate her position in IT.

She shared this experience with Marte, who also described herself as one of the best in the programming class at university, and who also experienced femininity as a disqualifying feature. People *"cannot quite identify me within a specific category"*, she said, because they don't understand *"how it is possible that I find it important with nicely manicured nails, and in addition I find programming both fun and easy"* (Marte). She did not fit into any of the predefined categories, neither as a woman, because she studied IT, nor as an IT student, because she accentuated her visual feminine features such as her manicured nails. The signs of femininity appeared to be incompatible, not only with the image of mastering IT, but, more specifically, with the idea of her being *interested* in and *enjoying* programming. The experiences of both Liv and Marte illustrate the entanglement of gender, competence, and interest in the construction of the norm of IT.

Reconstructions

In Chap. 3 we explored what had enabled, supported, and motivated the women to enter programs of IT, and found that gender was rarely a topic in those stories. This chapter has explored how the women experienced IT as gendered, and these narratives add to our understanding of the pathways as reflecting a gendered landscape, with bumps and potholes for women on their way to IT (Branch, 2016). Here we have seen women's marginality in spaces of IT constructed with reference to the stereotypical but still effective male storyline, putting women's non-conformity on the spot. Cultural images of women succeeding and thriving in IT had been scarce for most of them during their childhood and teenage years. Thus, the women's narratives highlight that most of them had not entered fields of IT because they felt invited; they had done so rather *despite* having little insight into IT and being conscious of a masculine image of IT that did not appear inviting. The women's perception changed as they entered spaces of IT, from a place occupied and defined by boys and men, to a place where also women could participate, and engage their "girl power" by building alliances and making women visible. The reconstructions of IT that followed opened the possibilities for women to *feel like insiders* in a space of IT they recognized as populated by *different types of people*, accepting *various types of interests*, and requiring *a multitude of skills and competences*.

Spaces and bodies affect each other in a two-way relationship, Puwar claims. On the one side, "specific bodies are associated with specific spaces", and on the other side, "spaces become marked as territories belonging to particular bodies" (Puwar, 2004, p. 141). While space invaders disturb the norm of IT, the space invader identity is also *productive*: by bringing their norm-disturbing features to the inside, women added counter-stereotypical examples, role models, and lived life to the narratives about what IT is and who can be considered an IT expert.

Those who fit in a space "do not see the tacit normativity of their own specific habitus, which is able to pass as neutral and universal", Puwar explains, while "those who attempt to *name* the particular—in terms of gender, race or class—in what passes as universal face the contortions of naming something that is ontologically denied" (Puwar, 2004, p. 131). It was not the formal barriers that challenged women's participation, but the historical and conceptual entanglement of gender, interest, and competence, where women failed on all three: having the wrong gender, lacking the background experience leading to a certain type of competence, which raised doubts about the depth of their interest. The way women bump into challenges is reflected in their initial doubt about their abilities to master IT and questioning whether they would fit. Their strategies for overcoming these barriers are, however, reflected in the alternative pathways, their emphasis of alternative competences as equally relevant, and in the efforts of claiming women's visibility. Defining their interest in much wider terms than only in terms of IT reflects the wide impact of digitalization and the relevance of digital technology, like the woman claiming her interest in IT based on her engagement in *the world*.

Different from dominance of the male storyline that had left little room for other (non-male) characters in the women's early perception of IT, their reconfigured understanding of IT included different types of people, a variety of competences and interests, and a multitude of uses for IT. The alternative competences and interests, such as engagement in society, sciences, languages, and creativity, were not uniformly gendered, but rather pointed to the universal challenges of modern digital societies. Thus, the work done by these women represents a collective effort to welcome values different from the male stereotype, but not thereby limited to women. The women's reconfiguration of IT opened the space of IT to "any(body)" (Puwar, 2004, p. 32) who did not conform to the male norm, illustrating that IT appeared quite elastic in the women's narratives. However, women who are not "the ideal occupants of [the] privileged positions" (Puwar, 2004, p. 11) do not just pass into spaces of IT without resistance. The women had experienced having their competence doubted, being perceived as less skilled due to signs of a female identity, feeling constantly under surveillance, and being scrutinized for whether or not they qualified. This narrative needs to be recognized as a cultural construction, illustrated by those foreign women who had been motivated to study IT *because* they were women. They illustrate that not all cultures see IT competence as a masculine field or an innate quality in men (Ensmenger, 2012). The masculine space of IT recognized in the Norwegian women's narratives is not a universal construction, but rather appears with different gendered configurations according to time and place (Blum et al., 2007; Trauth & Connolly, 2021). Although this suggests that the gendered images of IT could indeed be changed, the continuous underrepresentation of women in IT has often been interpreted as relying on girls' and women's lack of interest, ambitions or aptitude for IT (Stoet & Geary, 2018). This is an imprecise interpretation for at least two reasons. First, as we have seen above, women do express interest in IT; however, this often takes on a different form than the interest associated with young men. Second, seeking answers to the gender disparity in IT only in girls' and women's choices misses the importance of their environments, including potential supporters that should have cheered the young women on to IT, which is the topic of the next chapter.

CHAPTER 5

Girls Don't Walk Alone: Supporters' Investment in Welcoming Girls and Women into Fields of IT

Abstract Schools have an important role to play in making youth choose less gender-stereotypical educations. Schools can also play a significant role in opening the door to IT as a potential education for a wide group of young women. Through interviews with representatives from 12 Norwegian lower and upper secondary schools, this chapter explores how they consider their role in encouraging girls and women to become familiar with, and to consider studying, IT. Gender equality is a treasured value in Norwegian educational policy; however, schools have diverging views on what gender equality means in relation to IT, and also how to achieve it. The analysis demonstrates a lack of regulation and conformity in how schools address issues of motivating and encouraging girls to consider IT as a field of study.

Keywords School as recruitment channel • Recruitment strategies in schools • The paradox of interest • Strategy of doing nothing • Avoiding gender issues • Non-performative gender policy

INTRODUCTION: SHIFTING PERSPECTIVE TO SUPPORTERS

The previous chapters have demonstrated that most of the Norwegian women's perceptions of IT had been shaped by stereotypical images of male gamers and geeks, before enrolling in an IT degree at university. This

© The Author(s) 2024
H. G. Corneliussen, *Reconstructions of Gender and Information Technology*, https://doi.org/10.1007/978-981-99-5187-1_5

had made it difficult for young women to imagine themselves choosing IT, and they questioned whether they could succeed without a background involving gaming and programming that they associated with male students, such as other studies have found (Frieze & Quesenberry, 2019; Margolis & Fisher, 2002; Yates & Plagnol, 2022). Thus, there are several barriers and challenges that young women need to overcome before considering IT as a potential choice. Few of the women we have met had imagined studying IT during secondary school. A large group even started at university still unaware that IT would eventually become their chosen field. The turning point, the moment they started to think about IT as a potential study choice, could be identified as following an event or experience that had made them think differently about IT and how available it would be for them. The women's experiences demonstrate that they had needed some kind of input and insight before IT appeared as a relevant choice. Such input could come from many potential supporters, such as parents and friends, or from school (Gerson et al., 2022; Jacobs et al., 2017). However, with some exceptions, these groups played a minor role in the women's description of how they had come to study IT. Thus, while the analysis demonstrates the crucial role of input for women in choosing IT, it also documents that many women mainly recounted navigating this landscape alone.

From the previous chapters' focus on the women's experiences, this chapter shifts perspective to explore how schools approach the role of supporting and encouraging girls and women to become familiar with and consider IT. Nordic research has found family to be a more important source of insight into tech careers than early schooling for girls who made their choice early (Corneliussen, Seddighi, Simonsen, et al., 2021; Corneliussen, Seddighi, Urbaniak-Brekke, et al., 2021; Engström, 2018). School, however, represents a crucial arena for reaching a wider group of young women, in particular those without tech motivation from their home conditions. The survey among young women who had embarked on a pathway to technology education presented in Chap. 2 also identified a high proportion that reported about IT classes in upper secondary as an important motivation for pursuing a career in technology (see Corneliussen, Seddighi, Simonsen, et al., 2021). Considering the space invader challenge discussed in the previous chapter, this could indicate that experiences at school can have a significant role in preparing women to enter spaces of IT. But how do the schools consider the task as well as their own role in efforts to make girls and women consider IT as a future study and career path?

Interviews with teachers and counsellors from 12 lower and upper secondary schools across Norway were conducted in 2020 and 2021, as part of an evaluation of a national recruitment initiative for getting more women into technology.[1] The national *Girls and technology* campaign travelled across the country with a large show aimed at encouraging young women in lower and upper secondary school to pursue careers in technology (see Chap. 2). One key technique was to invite the young women to meet female role models in science, technology, and mathematics—some of the core disciplines leading to a career in technology. The campaign involved schools as organizers for sending the young women to these events. The schools that participated in the interviews had been part of this event, sending young women to the *Girls and technology* show, where only girls and women were invited to participate. The interviews explored how the schools responded to this as well as how the schools more generally approached the issue of gender disparity in technology educations and occupations (Corneliussen, Seddighi, Simonsen, et al., 2021).

Schools' Attitudes and Strategies for Supporting Women into IT

Norway and its Nordic neighbours are recognized as the most egalitarian countries in the world, according to international ratings (World Economic Forum, 2020b) and there is a high level of trust in gender equality as a common goal in the Nordic countries (Martinsson & Griffin, 2016; Teigen & Skjeie, 2017). Gender equality is a widely accepted value in Norway (Larsen et al., 2021), and it is, together with equal access to education, a fundamental principle established by the Norwegian *Education Act* and the *Equality and Anti-Discrimination Act*.[2] Working towards the achievement of gender equality in education and educational choices is thus part of the schools mission. According to the Norwegian *Education Act* for primary and secondary education, all pupils are entitled to support independent of traditional gender roles. Though schools have some tools to work with, such as the subject *educational choices* in lower secondary school, there is no coherent framework for guiding the schools' practices in detail, which leaves many decisions in this field to the individual schools.

The educators shared a general *agreement of gender equality as a norm* and a goal in education and in working life and never actually questioned the goal itself. Reflecting this, the school representatives also expressed sympathy to the goal of getting more women into fields of technology. Yet the interviews demonstrated diverse, even contradictory, ways that schools were engaged in questions regarding motivating and encouraging girls and young women to consider IT as a study choice. While there was some variation in terms of the schools' responses, there were also some distinct patterns that appeared, and I will discuss four of these below.

Hesitating to Focus on Gender Differences

One of the topics with diverging opinions between the school representatives concerned the issue of gender differences and gaps in education and occupations, and whether this was a topic for schools to engage in. Some schools involved issues of gender differences, in particular when discussing career choices: *"Our strategy is to talk about gender and untraditional educations and occupations in the subject called 'educational choice'. Here we talk about pathways to different occupations, and gender is a topic. Girls must dare to think unconventionally"*, a teacher from lower secondary school said. Another teacher from a lower secondary school provided the opposite viewpoint: *"We try to be gender neutral when we talk about professions and education, and we don't say anything about boys' professions or subjects"*. The last view was shared by most of the school representatives, who rather preferred gender-neutral ways of approaching questions and information about occupations and career choices. This also applied to upper secondary schools where they preferred to *"not focus specifically on gender"*, as one of the teachers explained.

Other studies of career guidance in Norwegian primary education suggest that such attitudes are widespread. While gender is considered "one of the major influences" on career choice for girls and boys (Mordal et al., 2020), many school counsellors find questions about gender difficult to engage in, and therefore often put them aside (Buland et al., 2020). The reluctance to problematize gender as an influence on career choice in school, however, indicates that the mere idea of initiatives targeting girls can be perceived as challenging. The interviews confirmed that most schools preferred recruitment initiatives for both boys and girls: *"We participate in science events, but they are targeting all pupils, not just girls"*, a teacher from upper secondary school explained. In addition, sending only girls to events happening during school time like the *Girls and technology* shows did, was not very popular. Som of the teachers suggested that it exaggerated gender as a difference and left the boys without a similarly engaging activity. The concept of targeting girls rather than focusing on all students thus seemed at odds with a core value of *sameness* in Norwegian schools (Corral-Granados et al., 2022), here interpreted as treating all pupils in a similar way. This, along with what appeared as a silent agreement of not exposing gender as a differentiating category in education and working life (Corneliussen, Seddighi, Simonsen, et al., 2021) in most of the schools, challenged the ability to engage in activities which targeted only girls to start with.

Distrusting Women's Interest in Technology

Can girls be recruited to IT? The question of whether initiatives for recruiting girls to technology could be successful at all, also seemed to split the school representatives into different camps. Some thought that such initiatives had a positive influence on girls and were quite certain that they had made more young women choose to study technology. This made the recruitment campaign targeting girls unique and therefore quite important as a resource offering up-to-date information about technology professions, one of the teachers admitted. Others rather assumed that any potential effects of the event would wear off quickly, while yet others doubted altogether that it had any effect at all:

> Unfortunately, the girls mostly listen to each other. Or they have an older sibling, and some have parents [in technology]—I think that is a major influence. The final decision, however, is often based on what their female friend does. (teacher, upper secondary school)

A common rhetoric among the teachers and counsellors was the recognition of the *unfortunate* situation that fewer girls than boys choose to study science, technology, and mathematics. This, however, often appeared together with a fatalist observation such as this: *"often the girls don't continue with sciences after upper secondary school, unfortunately. But that's just how it is"* (teacher, upper secondary). This rhetoric demonstrates that the teachers are certainly aware of the gender patterns in educational choices and describe them as *unfortunate*. However, this rhetoric also indicates that the teachers perceived this as something that was impossible to change. Consequently, the task of turning young women's interest towards technology was perceived as outside the school's control.

Attitudes to these questions among the relevant school personnel affected the schools' strategies and activities for dealing with the gender imbalance in technology. Few of the schools in this study had an active strategy for making girls more interested in technology:

> We have nothing special. In my opinion, it is the girls' choice. [...] There are quite equal opportunities [for men and women] in Norway. I therefore believe that girls and women must have a wish to do it. (teacher, upper secondary school)

The upper secondary school teacher confirms that the school does not have a specific strategy for encouraging girls to think about technology. They doubted any such effort was necessary, trusting that national gender equality policies over time have removed any barriers for girls and women to choose the career they want in Norway. Taking for granted that formal gender barriers have been cleared away, what is left appears to be girls' and women's *wish* to study technology.

Thus, while the school does not have an active response to the gender disparity in technology, the teacher's reflections suggest that the passive strategy can be understood as an intended support for women to make up their own opinion. Thus, the doubt about girls' and women's interest in technology, which the teacher shared with several of the school representatives, indicates an acceptance of the gender differences in educational choices as a reflection of girls and women pursuing the career they want.

While the educators had diverging views upon *whether* efforts to recruit girls to IT would work, their views upon *when* this recruitment could take place was even more divergent—also this affecting the schools' decisions to focus on this topic. One school representative recommended starting early, *"even at kindergarten"*, to make girls familiar with technology. Another suggested that lower secondary was *"too early"* to make girls interested in any specific occupation including technology, while yet another claimed that it was *"too late at upper secondary"*; by then, the women had already made some fundamental educational choices that opened some career paths while closing others. Paradoxically, when these two educators' viewpoints came together, they would in fact be right: *not* encouraging girls at lower secondary to consider choosing science, technology, and mathematics would not equip them to be considered a target group for arrangements such as *Girls and technology* at upper secondary.

Sorting Women by Interest

Similar thoughts were reflected in the way schools filtered who they invited to recruitment arrangements such as the national *Girls and technology* campaign. Lower secondary schools mostly involved girls from entire classes for the campaign events, where technology and related occupations were presented by female role models. However, some lower secondary and most upper secondary schools rather invited individual girls based on who they thought would be interested: "*as adviser I encourage girls who are good at maths and science*", one explained, while another school only involved "*girls who have chosen science. We can't reach all, only those who are interested*".

The logic by which some of the schools operated reflected a negative circle, where girls who had not explicitly expressed an interest in computers, technology, science or mathematic were not perceived as interested in technology. Thus, they were not always invited to learn about technology or to attend recruitment activities targeting girls, thus leaving them with less insight and less interest. The schools doubt that young women's study aspirations could be influenced by school more than by family and friends resulted in a paradox in which young women's greatest chance of receiving support and encouragement to learn about IT was if they were already considered to be interested in IT. Figure 5.1 illustrates the paradox of defining interest in IT as a prerequisite for being invited to learn about IT.

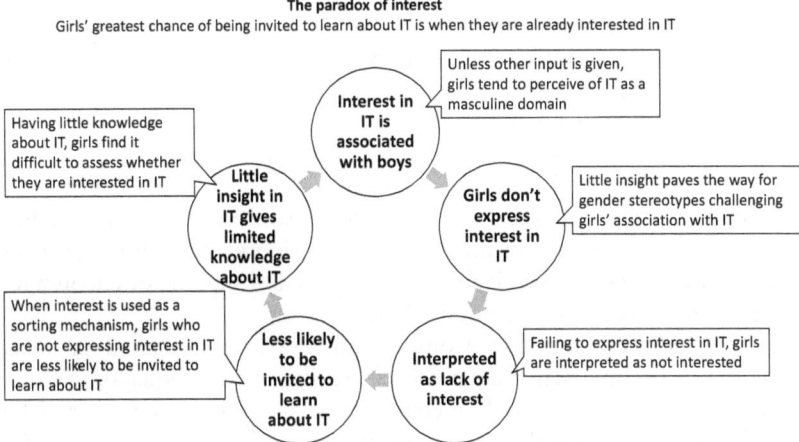

Fig. 5.1 The paradox of recognized interest as a prerequisite for being invited to learn about IT

Although all the schools participating in the study had sent young women to the *Girls and technology* events, they also had various arguments for *not* sending individual women or classes to the events, from too busy at school to challenging logistics. However, the logic expressed by some of the school representatives, of only inviting the girls and women who were assumed to already have an interest in technology, reduced the number of young women who were likely to participate in such recruitment initiatives. The schools' sorting of women by interest also reduced the possibility of seeing the more *dramatic effect* of recruitment initiatives and other events, identified when women who had never before imagined themselves studying or working with technology, experienced something that had put technology on the map as a potential study for them.

A BURDEN OF DOUBT PAIRED WITH POSTFEMINIST ASSUMPTIONS

Many of the educators shared a doubt about girls' and women's interest in computers and IT, reflecting the "burden of doubt" that often challenges space invaders in Puwar's vocabulary (2004). Furthermore, many were also doubtful that the women's study aspirations could be changed, echoing an essentialist understanding of gender as innate qualities of men and women, rather than seeing gender as a result of what we do and how it has been constructed through social relations (West & Zimmerman, 1987). One effect of this doubt was, as discussed above, that mainly girls already recognized as interested in technology would be invited to learn more about technology at schools. Thus, while gender stereotypes work as a barrier for young women to imagine themselves fitting into the male spaces of IT, educators add to these barriers when they are limiting invitations to learn about IT to women already considered interested. The women this is likely to affect the most are those who do not have family, friends, or other sources for motivation and insights into technology, thus making it even more difficult for this group to find support or encouragement, or to be recruited through school.

The educators' doubt in women's interest in IT resulted in weak attempts to become active agents for increasing gender equality in disciplines of technology. The lack of engagement, however, was not perceived as being in breach with the gender equality norm. The impression of gender equality already being installed in society was strong in the schools, reflecting a Nordic myth (Martinsson & Griffin, 2016) and a branding of

Norway as a superpower where gender equality is already in place (Larsen et al., 2021). This reflects a postfeminist attitude, which is not a branch of feminism but rather refers to an assumption that whatever structures in society which previously produced gender barriers have been removed and that any remaining gender inequality "can be accounted for by choices knowingly made by individuals" (Budgeon, 2015, p. 304; Corneliussen, 2021a). Thus, what is left, is individual choice. This changes the responsibility for gender inequality from policy and societal structures to individuals, in this case to women. Furthermore, it contributes to a perception of gender-typical career choices as a legitimate result of free choices and therefore as something that needs to be supported rather than resisted (Ellingsæter, 2014, p. 87). This describes the educators' responses quite well; their *regrets* but still *fatalistic* claims of trusting that the young women follow their hearts when making career choices. In a postfeminist perspective, women's underrepresentation in IT is thus interpreted as a result of girls and women *not wanting* to participate in fields of IT, and not as a result of, for instance, not inviting girls and young women to learn about technology. The postfeminist trust in gender equality already being achieved thus supports a rhetoric of not *forcing* women to change. Furthermore, it justifies the schools' weak efforts to create change or become active agents for gender equality interventions in fields of technology.

When Doing Nothing Appears to be the Right Thing to Do

The previous chapter illustrated that many Norwegian young women had shared a space invader experience of disturbing the norm in a masculine space of IT. However, norms producing excluding boundaries only become visible when they are challenged by bodies "out of place" (Puwar, 2004, p. 49). Gender boundaries have a tendency of remaining invisible for those "able to pass as neutral and universal" and who do not challenge the norm (Puwar, 2004, p. 131). The educators did not share the women's perspective and their experience of being *constructed out* (Puwar, 2004, p. 42) by the masculine norms of IT. The "myth of sameness", Puwar explains, rather makes gendered spaces such as IT appear as neutral and universal (2004, p. 131). This undermines charges of barriers or challenges affecting women more than men (Kaiser et al., 2013).

Gender equality can, in these examples, also be understood in terms of a different type of myth, that is, in Laclau and Mouffe's notion of a myth as something everybody apparently knows without, however, having agreed upon a precise definition (1985). While gender equality for some indicated a need to increase women's participation in IT, for others it translated into a freedom to choose, or what Charles and Bradley refer to as a liberal egalitarianism and "the right to choose poorly paid female labelled career paths" (2006, p. 195). This means that supporting women in whatever choices they make rather than trying to convince them of making less gender-conventional choices, is also perceived as supporting the goal of gender equality in the shape of the right to free choices (Corneliussen, 2021a).

For some of the school representatives, the low proportion of women studying IT seemed to confirm the assumption of women's low interest in IT. This worked to further reduce the school representatives' *perceived responsibility* for producing change. Interpreting this in the light of gender equality translated into free choice simultaneously reassured that they were not in conflict with the gender equality norm. This explains how schools' rather passive response to gender imbalance in IT was not necessarily perceived as a fault to support gender equality and not as a resistance to or disagreement with the goal, such as gender equality actions have faced in many examples (Bleijenbergh, 2018; Dick, 2004). Here it was rather interpreted as a support for the democratic qualities of a free and gender-egalitarian society (Corneliussen & Seddighi, 2020a). Thus, it made it seem like doing nothing was the right thing to do in the name of gender equality.

The Schools' Responses Versus the Women's Experiences

Above we have seen the school representatives describing the schools' responses to questions of gender disparity in technology. While there was certainly variation between the schools, four distinct patterns were apparent; firstly, hesitating to put gender on the agenda; secondly, distrusting women's interest in IT; thirdly, sorting women by interest before distributing invitations to learn more about technology; and finally, trusting the national level of gender equality to already have sorted the issue, thus making no response the right response. These four patterns are worth noting since similar patterns have been identified in previous research of schools (Mordal et al., 2020), computer clubs (Corneliussen & Prøitz, 2016), and IT sector responses to gender disparities in IT (Corneliussen & Seddighi, 2020a, 2020b). Furthermore, these attitudes, affecting the choices made by schools in these matters, have consequences for how girls and young women experience school as an arena for motivating them to think about IT as a relevant career path. Thus, the interesting questions here are: how do the attitudes and responses illustrated by the school representatives match with the experiences of the women we met in the previous chapters? How well are the schools' strategies aimed to support women into putting IT on their educational horizon?

Avoiding Gender Issues Reduces Schools' Role as Supporters for Women

Several of the schools adhered to a limited focus on gender differences as their preferred way of dealing with gender gaps in education and occupations. This response makes sense in light of schools' emphasis on activities that target boys and girls in similar ways, and in light of "sameness" as a core value for schools (Corral-Granados et al., 2022). The tendency of avoiding discussions about gender differences has also been identified in previous research as a general lack of focusing on gender equality at schools, and such questions are often put aside because they are difficult to engage in (Buland et al., 2020; NOU, 2019: 19), not only in STEM fields, but also in the female-dominated field of nursing (Lien, 2021). Furthermore, a similar strategy of avoiding putting words to challenges of recruiting women into IT jobs has also been identified in the IT sector. Here, issues of gender difference and gender equality involved notable challenges for IT companies with little knowledge about how to deal with such questions (Corneliussen & Seddighi, 2020a).

In a space invader vocabulary, the avoidance of bringing up issues relating to gender differences, can be interpreted as an endorsement of IT as a seemingly neutral space "that can be filled by any(body)" (Puwar, 2004, p. 32). However, as the women have shown us, they had experienced IT not as neutral, but rather as a masculine space defined by assumptions about young men's intense relationships with computers. Bodies that fit the norm can pass through gender boundaries without resistance (Ahmed, 2012); however, many women's experiences include uncertainty of whether they would fit and whether they could compete with men who had been playing with computers since their early teenage years (Margolis & Fisher, 2002; Yates & Plagnol, 2022)

One question that remains to be answered, is whether it would have mattered for the women if schools had focused on gender as a difference in IT in the classes. However, failing to consider how gender structures and stereotypes affected young women's attitudes to IT as a potential career path might have not only affected girls directly, but also worked to limit the schools' response to this challenge. Schools' avoidance of addressing gender differences thus seemed to reduce the schools' potential role as supporters for those women who never had thought about studying IT.

A Narrow Definition of Interest in IT Excludes Alternative Types of Interest

The tendency to distrust girls' and women's interest in IT developed into a pattern of sorting women by (recognized) interest in IT when deciding who to invite to events for learning more about technology. This suggests a doubt that women's interest for IT can be sparked at all, if it was not present already. This distrust echoes, for instance, studies documenting that very few girls compared to boys are aspiring to pursue a career in IT (Borgonovi et al., 2018), and few become interested through school (Alshahrani et al., 2018; Engström, 2018). The assumption of women's lower interest in IT has been identified across many other contexts (Sultan et al., 2019), such as family (Tænketanken DEA, 2019), leisure activities (Corneliussen & Prøitz, 2016), gaming (Dralega & Corneliussen, 2018; Sevin & Decamp, 2016), and in the IT sector (Corneliussen & Seddighi, 2020a). Research has even found that women themselves doubt that they have enough interest in IT compared to men (Margolis & Fisher, 2002).

Also this focus on interest in IT can be understood in light of the space invader vocabulary as a doubt about the motivation behind women's participation; a "burden of doubt" that questions their willingness and ability to participate (Puwar, 2004, p. 11). How well does this distrust in women's interest and the related sorting match with the women's own experiences? More importantly: how does it fulfil the support that women need for considering to study IT?

Putting interest in IT as a key qualifier for receiving support to think of IT as a career path fails to identify that many different topics and disciplines had led women to an interest in studying IT, from science and mathematics to arts and creative writing. Nor does it recognize that neither might an interest in IT be enough to lead women to choosing IT when they face other gender barriers (Corneliussen & Seddighi, 2022).

Furthermore, this attitude also fails to recognize one of the most common situations that the women reported in the previous chapters, that of lacking insight and thus only having had vague ideas about what a career in IT really involved. Few of the women recalled activities at school sparking their interest in IT. Thus, the negative circle of interest leading to excluding the young women who had not expressed an interest in IT reflects the women's experiences of weak support at school. This includes Lene, who pointed out the obvious, that *"it's no wonder that I didn't become interested in IT when there was nothing available"*, and Gro, who suggested she would have *"sat down and started programming right away"* if anyone at all had showed her the fascinating world of programming earlier. Remembering Ingrid's struggle of how to bridge the two seemingly contradictory positions of *being a girl* and *being interested in technology*, the interviews with the school representatives suggest that girls who had not been able to bridge these positions, were less likely to be invited to contexts where they could learn about and become familiar with technology. Furthermore, while a certain set of technology- and science-related interests were central for whom the schools would invite to initiatives such as the *Girls and technology* events, Ingrid and the other women challenged this by identifying a much wider set of interests that had led them on to an IT degree at university.

The negative circle of excluding girls who did not already express interest in technology from arenas for learning more about technology, undermines the effect of such initiatives. The evaluation of the *Girls and technology* campaign, for instance, illustrated a dramatic and abrupt change of direction for some of the girls who had no other sources for learning about technology as a career path. A similar effect was illustrated in the women's narratives presented in the previous chapters, with many examples of how different experiences had sparked their interest in technology in surprising and unexpected ways. These stories illustrate that young women without sufficient knowledge about technology also might find it difficult to express interest in technology, simply because they lack knowledge about what to express interest in. The previous chapter showed that women's perception of IT transformed from seeing it as a narrow space for male gamers and programming enthusiasts, to a more open space where "normal people", including women with a wider set of interests and competences, were also welcome. While the school representatives seemed, to a large degree, to share the initial gender-stereotypical notion of IT as a masculine field, they had not been through the same transformation. The women's many alternative ways of describing their interest for studying IT, including topics such as arts and creative writing, using technology for shaping society or an interest in the world, were not part of interests that the school representatives mentioned as qualifying for being considered interested in technology. Thus, sorting women by interest in technology undermines the possibility of raising this interest in women who had not yet quite made up their mind about technology.

When Gender Barriers are Cleared Away, Free Choice Remains

The last pattern was the burden of doubt paired with a postfeminist assumption, trusting the national gender equality policies to have cleared out any gender barriers. This leaves women's choices as the challenge, and the postfeminist *free choice argument* moves responsibility for women's underrepresentation in IT from a structural level to individual women (Corneliussen, 2021a). The consequence of this belief is reflected in few active interventions for raising young women's interest in IT. Instead, translating gender equality into a democratic right of free choice rather transforms this into a question of supporting women in whatever choice they make. The assumption that gender equality is already doing its job for the young women covers up the problematic structures and the experiences that many young women have of IT as a masculine space. This challenge is discernible in gender statistics which even finds this to be a more intense challenge in Norway than the OECD average (OECD, 2021; The Norwegian Universities and Colleges Admission Service, 2022). This suggests that the national branding of Norway as a superpower of gender equality (Larsen et al., 2021) has become a blockage for the very thing it names. The notion of gender equality itself does not do anything, but rather works as what Ahmed describes as a non-performative policy that appears to solve the issue merely by existing (2012). The assumption of equal opportunities that one of the teachers points to does not consider, for instance, women's struggle to bridge the seemingly contradictory positions of both being a girl and being interested in IT.

Putting the pieces together we find that recruitment events can have a major impact by making IT a visible and interesting study choice for young women. However, the passive responses of some schools are also reflected in how most of the women had rather experienced a lack of early support for thinking of IT as a relevant choice. This had shaped their pathways to IT, leading to a delayed entry into IT, penalty rounds, or just accidentally ending up in IT. Thus, the weak focus on these issues in the schools' strategies to a large degree reflects the women's experiences of having had to navigate the gendered landscape of IT mainly on their own, finding support in other places rather than through school.

Girls Still Doing IT for Themselves

Women do not operate in isolation when navigating the landscape of education, work, and career. Norwegian youth are affected by gender stereotypes when making their study and career choices (NOU, 2019: 19). Growing up in a culture where IT is associated with men made support, motivation, and insight into IT vital for the Norwegian women to consider IT as a relevant study choice. There is no doubt that gender equality is considered a treasured value in Norwegian schools. Thus, all the school representatives agreed with gender equality as a goal; however, what gender equality actually means, was less obvious. While all the participating schools had sent young women to the *Girls and technology* events, they had different attitudes towards the campaign and, more generally, to how to deal with gender disparities in educational choices. The gender perspective of doing gender underpinning the analysis here emphasizes that gender is constantly produced and reproduced in social interaction (West & Zimmerman, 1987). Fenstermaker and West's notion of "doing difference" points to how people are producing or "doing difference" between themselves and others, resulting in such differences appearing as natural, "as if the social order were merely a rational accommodation to 'natural differences' among social beings" (2002, p. 207). Here we find this illustrated in how some of the school representatives assume that a low proportion of girls signing up for IT classes or aiming for a career involving IT are reflections of natural gender differences, which further undermines the ability to work for change.

Although the number of interviews with schoolteachers and counsellors were limited and should therefore not be considered representative for all schools in Norway, the patterns reflecting the women's experiences indicate that they are not isolated examples. The tendency of schools avoiding topics relating to gender differences in education and working life also reflect previous research finding that gender equality is among the lowest-prioritized target areas in Norwegian schools (NOU, 2019: 19) and that questions of gender and career choice are often experienced as a challenging topic for career counsellors at schools (Buland et al., 2020).

"In Norwegian public schools, the teacher holds a role as a representative of society. The teacher converts the values and attitudes that the Norwegian state, through official documents, has defined as important in shaping well-functioning democratic citizens", Andresen explains (2020). In this case, however, the schools' autonomy and their varied responses to the task of encouraging girls and women to think about IT rather suggests an absence of clearly defined national strategies. The result is a lack of routines to ensure that schools become active agents for making IT appear as an equally relevant career path for young women as it does for many young men.

NOTES

1. This project was a collaboration that included my three colleagues Gilda Seddighi, Anna Maria Urbaniak-Brekke, and Morten Simonsen. See also the Norwegian report in Corneliussen, H. G., Seddighi, G., Simonsen, M., & Urbaniak-Brekke, A. M. (2021) and Corneliussen, H. G., Seddighi, G., Urbaniak-Brekke, A. M., & Simonsen, M. (2021) on some of the quantitative material from this project.
2. The Norwegian Education ACT: https://lovdata.no/dokument/NLE/lov/1998-07-17-61 and the Equality and Anti-Discrimination Act: https://lovdata.no/dokument/NLE/lov/2017-06-16-51 (accessed 23 October 2022).

CHAPTER 6

Gender Patterns, Equality Paradoxes, and Lessons for an Inclusive Digital Future

Abstract The aim of this book was to answer the question: what makes women enter fields of IT? This final chapter will sum up the lessons from studying the women's chronological pathways, space invader experiences, and reconstructions of IT, discussing the implications they might have for women, educational environments, and researchers. Learning points from barriers as well as turning points, and reconstructions that supported the women's entries into a university degree in IT, can become guidelines for an ecosystem of supporters interested in making a more gender-inclusive digital future. This involves a discussion of how this field is riddled with a gender equality paradox and a counter-productive postfeminist reaction that results in a non-performative gender equality norm.

Keywords Nordic gender equality paradox • Postfeminist myth • Reconstructing gender and IT • Reconstructed notion of interest • Ecosystem's responsibility • Gender-inclusive digital future

INTRODUCTION: A GLOBAL DIGITAL TRANSFORMATION

There is a global shortage of skills and a growing demand for IT professionals, calculated to a deficit of 20 million experts in key areas in the European Union (EU) by 2030 (European Commission, 2021b). Digitalization is transforming the labour market and working conditions,

© The Author(s) 2024
H. G. Corneliussen, *Reconstructions of Gender and Information Technology*, https://doi.org/10.1007/978-981-99-5187-1_6

and altering how we work. Emerging fields, such as artificial intelligence (AI), cybersecurity, robotics, and e-health, are expanding rapidly, taking new roles in new sectors and industries with accelerating speed (United Nations Industrial Development Organization, 2021). It is at present unclear who will be involved in this transformation, which is currently having gender-specific impacts (Barbieri et al., 2022). Despite differences between individual economies, the gender digital divide and differences between men's and women's access to technology and digital competence is reproduced across nations, irrespective of geographic location, economy, and the overall level of access to information and communication technologies (Sey & Hafkin, 2019). Differences in access to digital skills, education, training as well as opportunities in IT-related work have consequences for social difference, including access to social status, wealth, and power (Mariscal et al., 2019). This makes the gender disparity in IT a critical problem to solve, not only for women, but also for nations, economies, and companies (The European Institute for Gender Equality, 2017).

This global challenge of gender disparity in technology is notable also in the Nordic countries, which are simultaneously recognized for their gender egalitarian culture and their political regimes with a long history of gender equality policies (Hernes, 1987; Teigen & Skjeie, 2017). This book has illustrated that the cultures, images and imaginaries of tech professionals are still strongly affected by gender cultures and stereotypes in Norway, one of the Nordic "superpowers" of gender equality (Larsen et al., 2021).

In the context of these gender equality superpowers, the main goal of this book has been to develop our knowledge of how to produce a more gender-inclusive digital future, one in which it will be equally natural for girls as it currently is for boys, to opt for a career in technology. Starting from the realization that IT education is highly male-dominated in Norway, the aim was to learn from the women who did enter these fields of higher education. Chapter 2 contextualized the analysis by revisiting research literature on girls' and women's participation in IT. Chapter 3 presented the women's chronological pathways with a focus on the positive factors that had enabled them to enter higher IT education. Here gender was mainly unspoken, and yet gender was entangled in the women's experiences, which was elaborated in Chap. 4. Women's experiences in IT brought new perspectives and perceptions, while the women also themselves contributed new meanings, including a picture of IT as a more inclusive field. Realizing that women need insight and support to consider

IT a relevant career path, Chap. 5 analysed schools' attitudes and practices. Despite large variations, this showed a tendency for schools to passively trust that women made their own choices guided by the gender egalitarian culture of Norway.

Different from most studies that explore one arena or one transition, for instance from upper secondary to university, the women's chronological narratives, analysed using an explorative analytical model guided by grounded theory (Charmaz, 2006), helped to understand how women navigate the gendered landscape of IT in ways that do not always fit the traditional routes and models. Gender entered the Norwegian women's narratives when they recounted challenges and barriers, and when describing how they engaged to develop their belonging in masculine spaces of IT. A rather different story was narrated by the foreign women, of being encouraged to study IT because it was considered *suitable for women* in their home countries.

Research of, as well as solutions to, the gender gap in technology have often involved an idea of conventional routes to IT as shaped by an interest in IT, preferably developed early, followed by the "proper" choices in school that leads to a higher education in IT. The unconventional pathways that proved important for many of the women might not have been reserved for women alone. However, the women's association of IT with men greatly affected how they experienced the journey. This was evident in the women's accounts of feeling that they were disturbing the masculine norm, making them doubt about fitting in as well as experiencing the "burden of doubt" questioning their competence. The space invader metaphor (Puwar, 2004) captures the women's wish to participate, having the right to be there, some of them even excelling in core fields of IT competence, but still having a feeling of being not fully accepted within the masculine norm of IT.

The previous chapters illustrate how the conventional route to IT appears as a bumpy road that mainly fails to recruit young women, not the least because also the women's environments had internalized (Berger & Luckmann, [1966] 1991) a cultural image of IT as a masculine space (Puwar, 2004). Insight into how the schools understand and deal with the disparity of women in IT shed light on their ability to act as an arena for developing women's ambition to study IT. Some of the schools' postfeminist assumption (Budgeon, 2015) that *doing nothing is the right thing to do*, explains the weak role that schools took as well as the tendency for women having to find their own pathway and for "doing IT for themselves".

The effects of gender stereotypes remain one of the major challenges in Norway. Stereotypes still produce expectations to which activities that represent important routes to IT involves gaming and programming. Few of the women had this kind of hands-on experiences with technology. This led them to question whether they would fit in IT, and whether they could compete at all with the young men whom they assumed had already acquired competence in IT, especially programming, before higher education. These findings are nothing particularly new, nor are they unique for Norway (Margolis & Fisher, 2002; Vainionpää et al., 2019; Wong & Kemp, 2018; Yates & Plagnol, 2022). Thus, the challenges and barriers pointed to here are recognized across most of the western world where IT expertise and work are still highly male-dominated fields (Borgonovi et al., 2018; Mariscal et al., 2019). This, however, also raises the important question, of why things are not improving, not even in a gender egalitarian culture such as the Norwegian. Although we should not assume that reasons for gender gaps in IT are the same today as they were 20 or 30 years ago, it seems as if the world of digitalization and emerging technologies are constantly changing, while the gender patterns are stuck with women tuned to being underrepresented in numbers as well as in cultural images of IT.

While some of the stories about explicit gender discrimination in fields of IT shared by the women are certainly discouraging, we should perhaps be even more troubled by the stories of how doing nothing appears as the best solution, by actors who trust that the national gender equality norm has already done the job. Such stories fail to capture the importance of being invited and initiated in the secret masculine spaces of IT for the women, the turning points that put IT on their agenda.

The insights into the schools' responses to gender patterns in IT, along with their role in encouraging women to choose gender untraditional technology education discussed in previous chapters, give insights into how not only women should be a target group for initiatives aiming at fixing the disparity in IT. Other actors such as schools should also be included as target groups for such initiatives, as their attitudes, actions, or lack of action also shape the women's experiences of IT. It is vital to look beyond women's educational choices to find the necessary answers and solutions to the challenge of gender equality policy losing its guiding quality as it rather takes on the impression of being a description of society instead (Ahmed, 2012). This suggests that educational authorities hold a key role in facilitating and providing guidelines for improving the gender balance in fields of technology.

Here we have the possibility to learn from women who did succeed in finding their way to IT. Below I will sum up the main lessons from the women's space invader (Puwar, 2004) experiences, as they described not only barriers but also events representing turning points for putting IT on their agenda, and how their reconstructed images of IT opens for a more inclusive image of IT experts. Furthermore, the chapter will discuss how this field is riddled with paradoxes and a counter-productive postfeminist reaction that results in a non-performative (Ahmed, 2012) gender equality norm. Finally, the chapter will consider the women's experiences as a foundation for action points that a wider ecosystem can engage in to improve the gender disparity in IT.

Space Invader Lessons: Barriers, Turning Points, and Reconstructions

The previous chapters have explored how and why women find a route to IT, focusing on the events and people that contribute as positive drivers. However, we have also seen that spaces of IT appear as populated by men and shaped by a male storyline that contributes to expectations not only of who are participating, but also of what the "proper" routes to IT should include, such as gaming and programming as pre-study activities. The women's chronological narratives illustrated how most of them as teenagers with references to these assumptions, had considered IT a male space, and therefore their engagement in IT had been far from straightforward. The women's chronological journeys from childhood to IT education at university through this unfamiliar landscape were characterized by gendered barriers that had made them doubt their choice, turning points that had put them on a route to IT, and ways of reconstructing images of IT as they negotiated their own participation and belonging. The main findings relating to these three phases of experiencing barriers, turning points, and reconstructions are illustrated in Table 6.1.

Barriers The barriers identified here are related to the cultural and stereotypical gendering of IT. This shapes the women's space invader experiences of disturbing the masculine norm of IT. However, it also affects the women's environments and potential supporters, such as schools. The main barrier for early ambitions in the direction of IT is the young women's *lack of knowledge about IT*, which makes few of them consider IT as a relevant study choice at secondary school and when advancing to university. While IT is relatively invisible at school, once it is put on the agenda, it initially appears

Table 6.1 Space invader lessons identifying barriers, turning points, and reconstructions

---

BARRIERS

Lacking knowledge about IT
Lack of insight and knowledge about IT does not make IT appear a relevant study

Gender stereotypes presenting IT as a masculine space
Gender stereotypes have a major impact on young women's ambitions to enter fields of IT and make them question whether they will fit in

The invisibility of women in IT
Lacking female role models and images of women succeeding in IT make women question their belonging in IT

Lack of support
Many young women experience little encouragement and support in navigating the landscape of IT

Women face the "burden of doubt"
Women feel that they have to prove their competence in masculine spaces of IT

TURNING POINTS

Insight is key to making IT relevant
Events providing insight in IT or hands-on experience were crucial for making IT visible as a relevant study and career choice

Seeing IT as relevant for anything
Many women became interested in IT when seeing its relevance for nearly any field and occupation, opening for a wider set of background and interests

Seeing women as equal participants in IT
Meeting female role models was crucial for some of the young women's ability to imagine themselves in IT

Finding unconventional pathways to IT
Failing to be recruited through school, many women found unconventional pathways to IT, including the detour and a random entry

RECONSTRUCTIONS

Redefining interest
A wide set of disciplines, from mathematics and sciences to humanities and arts, society and the world, can lead to an interest in studying IT

Multi- and cross-disciplinary competence
Reflecting the unconventional pathways, detours, and redefined interests, many women develop a multi- and cross-disciplinary competence profile in IT

Claiming space and visibility for women
Women challenge the masculine stereotype of IT by claiming women's visibility and IT as an inclusive space

---

through gender stereotypes for most of the Norwegian women. Thus, *gender stereotypes presenting IT as a masculine space* further inhibit young women's ambition to enter fields of IT, making them question whether they will fit in and causing them to doubt their competence. Further adding challenges to imagining themselves working in IT was the *relative invisibility of women and female role models in IT.* The absence of images of women succeeding in IT made women question their belonging and, for some, this also created a feeling of being "alone" or being the first woman in the field. Most of the women, including those who did develop an early interest in IT, experienced *little support in building knowledge and interest,* and in finding the (best) route to IT. Absence can be difficult to identify, however the women's pathways are to a large degree shaped by absence; by their lack of insight and knowledge about IT, missing invitations to learn about IT, missing adequate guidance and lack of support, and of not being recruited when moving from lower to higher secondary, and when entering university. These gaps of insight and support impacted their pathways, resulting in the delayed entry or a detour that sent many women on a first round of studying something else before discovering IT as their preferred choice. For some this felt like a *penalty round* that they, quite literally, paid a high price for. Although women's experiences in IT gradually challenged and changed their perception of gender stereotypes, they then also faced gender barriers. More particularly, women faced *the "burden of doubt"* as "bodies out of place" who were disturbing the norm in masculine spaces of IT (Puwar, 2004). Even some of the most skilled women could sometimes feel that they had to prove their competence, simply because they had the wrong bodily cue for appearing as an expert in, for instance, programming.

Turning points The turning points identified also reflect how the women move through the gendered landscape of IT in specific ways. Lack of knowledge about IT was a major barrier, and this is reflected by the most important turning point common for all the women: *gaining insight in IT.* Hands-on experience, in particular with regard to programming, was crucial for making the women start thinking about IT as a relevant and interesting study for themselves. The women recalled this experience as a surprise in which they realized that programming was nothing like the boring spreadsheets they remembered from school. Programming was completely different from all other school subjects, as Anna explained, and therefore it was impossible to know from school experience whether or not you would be good at programming. While programming for most of the women still was the one area of IT where they would not take up the competition with men that they assumed to be more skilled, being initiated in the magic world of programming was also the one thing that several of the women recalled as the moment they were "hooked" and realized that studying IT was their highest dream. Thus, programming, often assumed to scare women away from IT (Denning, 2004), was one of the activities that most clearly had triggered many of the women's interest in IT, as a special and new competence promising an empowerment through new ways of producing and controlling technology.

Another important turning point for the women was the realization that *IT is relevant for nearly any field and occupation*, from being an artist to being interested in "the world". Struggling with the gendering of IT, most of the women also appreciated meeting female role models. *Seeing women as equal participants in IT* and role models helping them to deal with the cultural contradiction of being a girl and being interested in technology, represented important turning points. While few of the women had been recruited through the more conventional channels of education, most of them came *to higher IT education through less conventional pathways*, brought there by their interest in another field or discipline. They entered via a detour that for some was a necessary period of development, while for others it marked a penalty round, and a surprisingly large group of women came to IT as a result of random events. These alternative pathways allowed women to enter IT via less masculine spaces.

In common for the **turning points that** made the women start thinking about IT as a potential career was the input of new insight by seeing, hearing, or learning about technology. Some women had several sources and experienced a gradual development of interest, while others, particularly those who had never previously thought about studying IT, often recalled an event that resulted in a dramatic change as if a rocket had transported them to a new place. The events that had a particular strong effect such as recruitment initiatives targeting girls, illustrate the power of seeing female role models—women who not only succeed but also enjoy working with technology and who have solved the conflict between being a girl and pursuing an interest in technology. The power of being initiated in this new image of technology was illustrated by the dramatic change of direction for some of the women, from a more gender-traditional education to technology.

Reconstructions The women's initial space invader experiences of participating but not fully belonging, gradually developed into more of an insider position as they challenged the gendering of IT. Their reconstructions of IT reflect the positions they found available. Despite the fascination for programming expressed by many of the women, this remained a field marked by the male storyline of the young men who developed this skill at an early age. Most of the women rather *redefined interest from IT to an interest in other disciplines and fields* when defining their own engagement and belonging in IT. While mathematics and sciences were important for many of the women, probably because these disciplines were obligatory for many of the IT degrees, the women also included social science, humanities, and arts in the backgrounds that justified their entry into IT. The target for their interest was not limited to IT but was much wider, even including "the world", as one of the women symbolically described it. Reflecting the unconventional pathways that involved detours via other fields, a notable group of women had their first university experience in a field other than technology. Ranging from life sciences to social science, economy, humanities, and the arts, this meant that many of the women developed *a multi- and cross-disciplinary competence profile in IT*. Finally, the women also challenged the hegemonic cultural images of IT experts by actively *claiming space and making themselves visible as women* in spaces of IT. This contributed to reconstructing IT from a narrow space associated with men to a more open and inclusive space, often reflecting the women's experiences at university where they found IT environments that included "normal people" and "people like me". The women's narratives guide a reconstruction of IT in which traditional images of IT competence as gendered are challenged by their translation of a wider set of competences into relevant backgrounds for IT.

The women's pathways illustrate that taking for granted elements such as early interest in IT, technology-related leisure activities, or even school as an important recruitment arena, risk overlooking other factors guiding women towards IT. This risk also applies to research that focus on interest, self-efficacy, and value related to technology alone, thus failing to see the translations and reconstructions that were vital for women's participation in IT.

The women's narratives also illustrate that women are not alone in navigating the gendered landscape of IT. While the women described a series of barriers for developing an early interest in IT, the school representatives had a more ambivalent understanding of girls' relationship with technology, similar to an ambivalent attitude to women in IT also found among representatives for IT organizations (Corneliussen & Seddighi, 2020a). While fully agreeing with gender equality and increasing women's participation in fields of IT as a goal, they observed that women made other choices and thus started to doubt women's interest. In the case of some schools this resulted in a strategy of sorting girls by (assumed) interest in IT, while the same view sometimes developed into a doubt about women's competence in the IT sector (Corneliussen & Seddighi, 2020a), reflecting the power of the hooded-gamer storyline. For schools as well as IT employers, the tendency to assume that these patterns reflected girls' and women's choices made *doing nothing* to recruit girls and women appear *the right thing to do*.

Lacking support, the women had therefore themselves been instrumental in finding their way into IT. They also had an important role in making IT a space more welcoming for women after entering higher education. Many of them had experienced being one among few or perhaps even the only woman in an IT class or at a conference. Such situations threaten to make women appear as a "token" (Kanter, [1977] 1993), making them visible as a woman while facing the "burden of doubt" as a professional (Faulkner, 2009; Puwar, 2004). The narratives that the women shared, however, rather showed their energy put into turning this into a positive situation; challenging the masculine space of IT by raising their hand to claim women's visibility in male-dominated spaces of IT, raising their voice to make sure that everybody realized that women were present. These actions were triggered by a space invader experience of feeling like an insider but still facing exclusion as members of a group not representing the bodily norm (Puwar, 2004). When challenging the disciplining power of the gender norm, they recognized this not only as an act on behalf of themselves, but also as a strategy for making women as a group visible and as an act of empowering women in technology (Corneliussen, 2021b).

Paradoxes, Postfeminism, and a Non-performative Policy

The women's narratives demonstrate that being a woman had consequences for how they perceived and approached IT. These consequences were not experienced in any random place, but in the midst of a culture recognized as one of the most gender egalitarian in the world (World Economic Forum, 2020b), and in a nation with a high level of commitment to gender equality (Larsen et al., 2021; Teigen & Skjeie, 2017). When considering the women's experiences of IT and the school representatives' reflections about their role in supporting women to study IT, there are several contradictions appearing, some of which are captured in the notion of a gender equality paradox. The paradox points to the mismatch between assumptions of gender equality already in place in society and continuous gender segregation in education and working life, reflected, for instance, in the low level of gender diversity in the tech workforce (Ellingsæter, 2014; Minelgaite et al., 2020). In international studies this has been labelled a *Nordic paradox*, pointing to how the recognized gender-egalitarian and affluent Nordic countries have a more extreme gender segregation than countries recognized as less gender egalitarian and less affluent (Chow & Charles, 2019; Minelgaite et al., 2020; Stoet & Geary, 2018).

As I have discussed earlier (Corneliussen, 2021a), international analysis of the Nordic gender equality paradox often rely on a particular rhetoric such as using the national gender equality regime as a reference point for analysing individuals' choices. Where the nation succeeds in proving a high level of gender equality, it appears as if women fail to exploit the same freedom of gender equality (Charles & Bradley, 2006). Furthermore, women's gender traditional career choices have been interpreted as a reflection of the national affluence giving less economic strain and therefore leading to lower career ambitions compared to women in less affluent countries (Stoet & Geary, 2018). Since the national gender equality is already seen to be in place, this triggers a postfeminist notion of gender barriers already being removed, thus suggesting that free choices, rather than societal structures, explain women's continuous gender traditional study and career choices (Budgeon, 2015; Corneliussen, 2021a). My suggestion here is by no means to do away with gender equality; however, the rhetoric of the gender equality paradox is problematic for several reasons. It simplifies the situation and produces an uneven analytical frame measuring the nation already celebrated as a *superpower* of gender equality (Larsen et al., 2021) against individuals' choices, which then appear to be responsible for the continuous gender segregation. Furthermore, the rhetoric risks to make a causal connection between facts that exist together but may not necessarily explain each other, such as national affluence and women's career choices (Corneliussen, 2021a). Similar to findings in Sweden, the women studying IT are a rather homogenous group with a high degree of social and educational capital (Engström, 2018) with education to at least degree level. In addition, most of them also have highly educated parents. Furthermore, the high proportion of women students in higher education, including in high-status disciplines (The Norwegian Universities and Colleges Admission Service, 2022), does not support an image of Norwegian (affluent) women having less career ambition.

The rhetoric of the gender equality paradox was echoed in some of the views expressed by school representatives. The schools supported the ideal of gender equality; however, one of the characteristics revealed in the interviews was the schools' divergent views. Thus, while some assumed that girls could become interested in IT, others rather believed that turning girls interest towards IT was an impossible task. This resulted in divergent views on whether or not to focus on gender, and conflicting perceptions of the school's role and responsibility to intervene in girls' and women's career choices. Some schools' practice of sorting girls by interest had the paradoxical consequence that girls' and young women's greatest chance of receiving support and encouragement to learn about IT was if they were already considered to be interested in IT.

The women's narratives suggest a different experience, not only of gender stereotypes as challenging their participation, but also of not recalling being invited to learn about IT at secondary school. Several of the women suggested that they could have been recruited to IT sooner if only they had been invited to learn about it earlier. The mismatch between the women's and some of the schools' interpretations of the situation demonstrate that women's choice of pursuing a career in IT should not be considered an isolated act. Rather, it needs to be understood in the context of how, among other things, some schools hesitated to focus on gender differences and the resulting negative circle made it less likely for girls to become initiated in IT (see Fig. 5.1). Furthermore, the conflicting views, which become visible in the different interpretations of the situation by women and some of the school representatives, illustrate a tendency for gender equality to change from being a political goal to becoming a *description of society*. This made it less important to deal with gender inequalities, thus affecting schools as potential supporters. A similar effect was found among IT employers that also questioned the possibility of making change, which affected their willingness and ability to address existing gender inequalities in IT jobs (Corneliussen & Seddighi, 2020a). In these examples the postfeminist assumption of gender barriers being removed (Budgeon, 2015) translates gender equality into freedom, including the notion of free choice that neither should nor could be forced (Snickare & Holter, 2021). This way it appears more important to support and protect the choices that girls and women make regarding their careers, even gender-traditional choices, than to try changing them (Corneliussen, 2021a; Ellingsæter, 2014). Thus, the passive responses found both among schools and in other contexts of technology (Corneliussen & Prøitz,

2016; Corneliussen & Seddighi, 2020a) illustrate how the postfeminist myth has become performative in the shape of a narrative that makes it possible for gender differences to be ignored, even in the context of the gender-egalitarian Nordic countries. The passive responses to gender differences can thus coexist in harmony with the widely accepted gender equality norm (Corneliussen & Seddighi, 2020a).

The conflicting perspectives suggest that the gender equality paradox cannot be explained in the light of girls and women's career choices alone. Instead, a wider context, including other groups' attitudes and actions that affect the young women's career choices, also need to be considered as a part of the explanation. The evidence in the previous chapters suggest that it is not women's low ambitions or lack of interest in IT that leads to their low level of participation in IT, but rather the stereotypes and lack of support leading their pathways towards detours and penalty rounds.

The discrepancy between the gender equality ideal and a reality of continuous gender segregation has made researchers ask if gender equality is an illusion (Sund, 2015). Others have pointed to the ironic effect of gender equality measures (Kaiser et al., 2013) that, when simply by existing, make it appear as if the problem is solved (Ahmed, 2012). The ironic effect makes it even more difficult to intervene when facing structures and practices that continue to reproduce gender inequality (Kaiser et al., 2013). This is one of the mechanisms producing a gap between well-meaning gender equality policies on the one side, and practices that are inhibited and constrained by a myth of gender equality already in place, on the other side. This inextricable and endless circle is hard to break because the policy itself seems to feed the myth. This is illustrated, for instance, in the schoolteacher's assumption that interventions to make girls interested in technology are superfluous since the national situation of equal opportunities in the end makes this a question of girls' preferences to pursue their free choice. It reflects what Ahmed calls a *non-performative* gender equality policy, which, simply by naming the goal, appears to have solved the challenge (2012). A new paradox appears, in which the national gender equality norm has lost its disciplining power, while the postfeminist myth has gained the power to guide schoolteachers as well as others (Corneliussen & Seddighi, 2020a).

The discrepancy between gender equality policy and practice has led to many contradictions for women in the Nordic countries, where, on the one hand, there is a strong public discourse emphasizing the importance of increasing women's participation in fields of technology, while, on the other hand, women remain a minority in both traditional and new emerging tech-focused work environments (Griffin, 2022). The gap between policy and practice has been identified as a major factor contributing to ongoing gender inequality in the Nordic countries (Griffin & Vehviläinen, 2021).

RECONSTRUCTING MOTIVATION WITH INTEREST AS KEY

Interest has been identified as a key for recruitment to IT. Above we have seen examples of schools relying on interest in IT as a prerequisite for being invited to learn about IT. In research, interest has also been an important feature for evaluating women's motivation to study IT (Cheryan et al., 2015; Lang et al., 2020; Master et al., 2016). Furthermore, interest is often seen in relation to ability, belief and how well women expect to master the core tasks, the value they assign those tasks, and whether gender stereotypes affect women's sense of belonging (Eccles, 2009; Master & Meltzoff, 2020; Sáinz & Eccles, 2012). In a study of identity expression threat, referring to a concern about inconsistency between gender identity and the identity expressed by study choice, Cheryan and colleagues found that this threat made it more likely for women than for men to *downplay their interest* in computer science (2019). The women's narratives, to a large degree, confirm theories of women downplaying the traditional interest in IT activities. However, instead of describing a weak or even missing interest for technology, the women's narratives point towards a very different understanding of what interest in technology can be.

Exploring the narratives through the metaphor of the space invader highlights how the ways in which the women reconfigure the field reflect a response to the masculine norm of IT, including their redefinition of relevant interests for engaging in IT. By relating to a wider set of disciplines the women have expanded the target from a narrow interest in IT to nearly any discipline. While the space invader experiences highlighted how the women came to disturb the masculine norm of IT, the women identified other types of background competence and interest. This new landscape leading to IT, one in which gender stereotypes are less exclusive and limiting, opens the possibilities for a wider set of backgrounds having relevance for IT, and also makes it possible to target a wider set of potential students.

Science, technology, engineering, and mathematics (STEM) fields are often treated as one in discussions of gender disparities and recruitment.

The findings here, however, suggest that there is something quite different going on when women make choices to enter fields of IT compared to other STEM subjects that they are familiar with from school. Vrieler and Salminen-Karlsson's study of computer science teaching and learning environments suggest that this has to do with the different "types of capital that are considered legitimate" in computer science compared to natural sciences (Vrieler & Salminen-Karlsson, 2022, p. 44). The women's chronological narratives confirm that few of them had access to the types of capital necessary for putting them early on a route to IT. This also helps to explain the importance of the wider set of backgrounds engaged in the women's narratives. Thus, instead of comparing themselves to the masculine stereotype portrayed through the hooded gamer-storyline, the women described their interest in IT through a wide set of disciplines and engagements, not all of them directly related to technology, and few referring to early interest in IT, and even less to gaming. Establishing their competence and interest in a different field, a discipline they already knew, allowed the women to hold on to something safe and recognizable, while entering the unfamiliar field of IT. Here they could also establish their ability belief and self-efficacy in a field in which they already had a high level of confidence. Although many of the women found their safe platform within mathematics or sciences, it was not limited to STEM disciplines, but also included humanities, economy, law, life sciences, and the arts, all of which were used for declaring their interest in studying IT. Most of the women expressed an interest in technology as being applicable to nearly any aspect of work and life. Thus, Ingrid had become interested in studying IT when she realized that she could "work with society, but also with technology", and Marte had found that programming "ticks all her boxes" since she wanted to become an author, and Ellen saw her interest in IT as an interest in the world, because *"technology is a large part of the world"*.

The women's reconstructed notion of interest that leads to studying IT highlights how a narrow perspective of interest tied to IT risks excluding these wider notions of interest. As Marte reminds us, a wider definition of interest, such as her reference to creative writing, does not exclude an interest in core fields of IT such as programming. Another characteristic for the women was that many of them had initially chosen a more gender-traditional education before experiencing this moment of change. A large group of the Norwegian women had started, or even completed, a higher degree before entering an IT degree. This suggests an explanation to a tendency for women working in IT to have *more education* than men working in IT, yet at the same time *less education in IT* than their male colleagues (BCS, 2019). This also illustrates a tendency for women to develop a hybrid competence, where they bring their competence in another field into IT. The multi- and cross-disciplinary competence profiles that women acquire as a consequence of pathways involving alternative platforms and detours, can be a good fit for the competence needed in the ongoing digitalization. Emerging fields such as AI, e-health, robotics, and cybersecurity also increase the need for cross-disciplinary knowledge in the IT sector (BCS, 2019). The women identifying their interest in society and in the world should be a good fit for fields such as AI that is no longer isolated to problems explored in computer labs, but have rather found their way into all corners of society (Castañeda-Navarrete et al., 2023; United Nations Industrial Development Organization, 2021). Furthermore, the lack of diversity among IT experts is recognized as one factor producing gender bias in, for instance, AI algorithms (Barbieri et al., 2022; Corneliussen et al., 2023). A more diverse IT workforce is necessary to ensure that the technology can reflect the needs of people across society (Palmer, 2021). The multi- and cross-disciplinary competence that the women exemplify can be of high value in the emerging tech fields that will need a huge number of professionals in the years to come.

Herein also lies a large potential for recruiting girls and women to IT and new emerging fields of tech work, as the women's narratives demonstrate that they can be invited to IT based on nearly any field of interest. Though, as women still experience gender barriers to their participation in technology, this means that interest in IT alone is neither a guarantee for making women pursue a career in IT, nor is it a requirement for doing so. This has consequences for strategies for recruiting girls and women. It also has consequences for researchers, as limiting the study to a narrow interest in IT and hegemonic routes to studying technology risks overlooking the space invader experiences that make many women move through the gendered landscape of technology in unexpected and unconventional ways.

DOING IT TOGETHER WITH THE ECOSYSTEM

Research has provided ample evidence of how relevant actors, including friends, family, school, and IT companies, affect girls' and women's study and career choices (Eccles, 2015; Lang et al., 2020; Wong & Kemp, 2018). The main challenge today, however, seems to be that many of these potential supporters renounce their responsibility, such as the professor we met in Chap. 1, suggesting that there is nothing they can do, and the schoolteacher in Chap. 5 claiming that girls make their choices based on girlfriends, doubting that their career preferences can be changed, or the IT professional suggesting that perhaps it is not so important to recruit women if they are not interested in IT (Corneliussen & Seddighi, 2020a). The study involving schools illustrated a lack of coherent responses to issues of gender differences in technology. This is not unique to Norway, however. Despite the recognition of how gender stereotypes affect youths' career choices across Europe, education institutions struggle "when implementing inclusion policies", the European Commission claim, due to a "lack of clear guidelines" for how to deal with these challenges (2021a, p. 29).

While recommendations for inclusion activities and measures have been provided for decades, and in a multitude of ways and formats, on national and international arenas (Castañeda-Navarrete et al., 2023; Pawluczuk et al., 2021), the evidence shared here and in other studies (Griffin, 2022) suggest that the main challenge today is the gap between policy and practice. The women's narratives, as well as schools and other actors' attitudes to the gender disparity in IT, point to how the potential supporters— those who should have been first in line to motivate girls and women— have doubt about women's interest in IT. Contrary to the passive strategy we have seen here, research has demonstrated that reaching goals of gender equality requires long-term and systematic engagement and investment in the goals of gender diversity for companies (Dixon-Fyle et al., 2020) and for academia (Lagesen et al., 2021). Schools can also become a resource for making young people choose non-traditional educations; however, this also requires a comprehensive and targeted engagement (Reisel et al., 2019).

The concept of *ecosystem* has been used to emphasize that a diverse set of actors such as family, educators, policymakers, and others are needed to develop an inclusive environment for recruiting girls into male-dominated STEM disciplines (Sammet & Kekelis, 2016; Traphagen & Traill, 2014). Involving a wider set of actors from across sectors and industries can support the need to present a wide set of role models, tasks, and values related to technology (Cheryan et al., 2013).

Thus, the challenge at hand is not about changing girls, but rather about strengthening the engagement and investment in these issues from a wide range of actors, from education to the private and public sectors, that can contribute to creating more diverse and inclusive images of what IT is, what it is used for, and who the IT experts are (Corneliussen & Seddighi, 2019; Talks et al., 2019). Thus, the ecosystem could change the situation that is reflected in the experiences of the women having to identify the pathway to IT on their own, to a situation where girls and women are invited to "doing IT together" with supportive actors in their ecosystem. The barriers, turning points, and reconstructions identified in the women's chronological narratives can help to identify some basic guidelines for developing a more gender-inclusive future.

Learning points from the women's experiences of *barriers* include the importance of inviting girls to learn about IT. Provide them with less gender-stereotypical narratives of technology. Encourage and support them (even if they do not initially express interest). Promote female role models and images of women in technology. Make sure to fight gender bias in tech education and work.

Important lessons from the *turning points* suggest again that invitation to learn about IT is key, hands-on experiences in particular, and that filtering women based on an assumed interest in IT will filter out potential targets. Facilitating images of IT as relevant for nearly anything can reach a wider group of women, and probably a wider group of men as well. Promoting images of women who not only succeed but also enjoy technology can make the field more welcoming for girls. Recognizing that it is never too early and never too late to recruit women to IT means that all relevant actors have a job to do.

Learning points from women's *reconstructions* also involve the varied set of competences they bring to IT, as this reflects multi- and cross-disciplinary competences that will be needed in the fields of emerging technologies. Finally, the women's participation in IT support the development and promotion of images of IT experts as a more diverse and inclusive group, disconnected from the masculine stereotype.

The postfeminist challenge of not seeing the need for action is still the biggest threat to future gender equality in IT. With this in mind, the list below offers six rules of thumb for any actors within young women's tech-related ecosystem:

- Do not underestimate young women's fascination for technology, however, do not overestimate their insight.
- Do not underestimate the effect of inviting, encouraging, and supporting women to become familiar with IT.

- Not doing anything different from before will not produce change.
- Claiming not to make any difference between the genders (being "gender-blind") will often mean that gender-discriminating practices can continue undetected.
- Claiming that there are no women applying to IT jobs indicates that you did not look where women are.
- Claiming that it is someone else's job to make girls and women interested in IT is true, but it is also your (and everybody's) responsibility.

THE INTERNATIONAL PICTURE

While the analysis here was based on studies performed in Norway between 2018 and 2021, they reflect challenges recognized in large parts of the world. The underrepresentation of women in technology is found across the western world (Barbieri et al., 2020; Eurostat, 2021c); however, it is not universal. More women find their way to computing fields in particular in Asian countries and former Soviet states (Charles & Bradley, 2006).

Using Norway as an example has provided insight into how a recognized gender-egalitarian culture (Teigen & Skjeie, 2017; World Economic Forum, 2020b) still struggles with a continuous gender imbalance in IT education and work. Women's underrepresentation in IT, as well as most of the challenges described by the women in this book, have also been observed in other western countries (Arnold et al., 2021; Holtzblatt & Marsden, 2022). Gender stereotypes, gendered cultures and structures have been identified as barriers for women's participation across the western world (Chavatzia, 2017; Cheryan et al., 2015; Cohoon & Aspray, 2006; Frieze & Quesenberry, 2015; Frieze & Quesenberry, 2019; Turkle, 1988; Wajcman, 2004; Watts, 2009). The challenge of identifying and developing girls' and young women's early interest in and ambition to study technology has also been identified (Borgonovi et al., 2018; Master et al., 2016; Master & Meltzoff, 2020; Yates & Plagnol, 2022), and the same is the case with women's fear about facing more skilled male students (Margolis & Fisher, 2002; Yates & Plagnol, 2022). A lack of support and school being experienced as an irrelevant arena for developing an ambition to study IT has been demonstrated in other Nordic countries (Engström, 2018) and also in the UK (Alshahrani et al., 2018). The low proportion of women working in IT has resulted in a lack of female role models in most western countries (Cheryan et al., 2009; González-Pérez et al., 2020), which has made it difficult for many women to identify with the field (Lang et al., 2020). Women's tendency to have varied educational backgrounds when entering fields of IT has also been identified in the UK (BCS, 2019). Among the less-documented features internationally is the wide set of interests that the women's reconstructed image of IT involved, although this might also rely on a trend to focus on a narrow interest in STEM disciplines when exploring women's motivation. One challenge identified internationally is related to women's larger part of care work and challenges relating to their work–life balance (Barbieri et al., 2018), which did not appear as a challenge in the Norwegian women's chrono-logical narratives. Furthermore, the Norwegian women did not report on equally negative experiences as Michell et al. found in Australia, of being "chased away from [computer science] as part of a border protection cam-paign by some males" (Michell et al., 2017). Thus, a consideration of the international situation of women experiencing similar barriers suggests that the knowledge produced in this book can provide some answers to this mystery, not only in the case of the Nordic countries but also across the western world.

Closing Reflections: The Untapped Potential

Women are not a homogenous group. They have different backgrounds, interests, and histories, and they find different pathways to IT education. Women's relationships with IT are not similar, and they are not excluded nor included in exactly the same way (Dee, 2021). The many different IT disciplines represented among the women are not uniform either. But there is still a consistent pattern showing that most young women grow-ing up in Norway do not think of IT as a potential educational option in the transition from secondary school to higher education. This remains the pattern until something happens and IT is put on their horizon as an option also for women; also them. Most of the women had not chosen IT education because they were encouraged by someone, but rather despite the many small and large barriers, including the lack of support that they had encountered. These included their association of IT with boys and men, or, alternatively, as a boring school subject; knowing little about IT educations and professions, expecting them to be populated by men that they assumed would have more knowledge about IT than them-selves. This even made them question whether or not they needed IT skills before applying to an IT degree. These were amongst the things that made them doubt they would fit in a world where "you only see men". To top the list of challenges, few of the women recalled having been encouraged or motivated by people around them or at school, which also reflect some of the schools' hesitancy to focus on gender differences. As the list of barriers is based on the experiences of women who already studied or worked with IT, there are reasons to assume that it would look at least as gloomy among women who never chose to study IT. Thus, this suggests that the right question might not be why there are few women in IT, but rather how women still find their way into IT education despite all these barriers.

In the end, the mystery that motivated this book, of how to create efficient and lasting initiatives to increase women's participation in fields of IT, appears to be less about women's lack of interest in studying IT and more to do with their experiences when traversing the gendered landscape of technology. Furthermore, in the larger picture it has more to do with gender equality having turned into a myth and a description of society, and a non-performative policy that feeds passivity, rather than women's lack of career ambition. The suggestion here is not to take gender equality off the political map, but rather to identify ways of making gender equality become an operative value that makes schools and other potential supporters put in an effort in making girls and women see fields of IT as being as inviting for them as it is for boys and men. And it is not really a mystery how to make IT more welcoming for women, if we are to believe some of the women who contributed their chronological narratives to this book, such as Berit suggesting that "it is really just about advertising it, having proper information about what it is, and to mention that it is not as difficult as many may think". This, however, applies to the Norwegian women. Those foreign women who had been encouraged to study IT because it was suitable for women, contribute to underlining the cultural construction of the Norwegian women's experiences. On a positive note, this suggests that the gender–technology relation can indeed be different. On a more negative note, even the women who had grown up in cultures which encouraged young women to associate themselves with a career in IT started to question their belonging in IT after meeting the male-dominated IT departments at Norwegian universities.

The many detours, penalty rounds, and random events are examples of many different routes that have led women into IT education. Despite not initially recognizing an interest in IT, many of the women described how a random event had fuelled their fascination, great joy, and becoming "hooked", once they started learning more about IT. This not only points to how insight is key, but also suggests a large pool of untapped potential for recruiting women to IT studies. The greatest loss is likely among those women who never considered studying IT because they never acquired the insight that made them able to bridge the seemingly contradictory positions of being a girl and also being interested in IT. One notable challenge is that the experiences that made the women change their perception of IT as a more open space do not seem to have much retrospective power: these insights had not been available for the women before they entered university. Paradoxically, for many of the women it was the pathways that took them on a detour that made them able to identify IT as their dream education. For some of the women, the detour was important in terms of preparing them for entering what they perceived as a masculine space of IT, such as Signe, who thought that she was not tough enough to do that when she was still a teenager. For others, the detour was rather a penalty round, an extra round they had to go because they had not been initiated in the magical world of IT during secondary school.

Are we now any closer to solving the mystery of how to increase women's participation in fields of IT? The women's message is clear: cultural images of gender, including stereotypes, matter for their ability to imagine themselves in a career in technology. Reviewed evidence suggests that in the current situation of IT being still strongly associated with men and perceived as a masculine space, girls and women need support and input for starting to see IT as a place where they can participate and thrive. Thus, while IT continues to be dressed up in gender stereotypes, targeting girls and women will remain important because traversing the gendered landscape of IT involves different challenges for women and men (Sultan et al., 2019). It is not unusual to assume that gender segregation in educational choices is a result of individual choices (Snickare & Holter, 2021); however, it is necessary to recognize the underrepresentation of women in IT as a structural problem that needs structural solutions (Ahmed, 2012). One route to a more gender-inclusive digital future involves an ecosystem of supporters for young women. The ecosystem, however, also needs support and better guidelines for becoming active champions for gender equality in fields of technology (European Commission, 2021a).

Some of the most interesting insights developed in the studies reported here have come from the encounters with women who never had imagined to study IT and still ended up there, such as Gro, who made a blind choice and now holds a PhD in IT. Reflecting their gendered journey of coming into spaces of IT as outsiders—lacking the expected background associated with boys and men, several of these women found support in character of Pippi Longstocking, the strongest girl in the world, from the children books by Astrid Lindgren. Pippi was herself an outsider who was disturbing the social norms guiding other children her age, finding most ordinary activities quite unfamiliar. However, she was never afraid of taking on new challenges, and it is this courage the women evoke when quoting the motto reflecting Pippi's take on life: "I have never tried that before, so I think I should definitely be able to do that".

There are still many unanswered questions of how to create a more gender-inclusive digital future. This book provides some answers, pointing to the responsibility of a wide ecosystem of parents, educators, policymakers, IT specialists and others for supporting, inviting, and encouraging girls and women to think of IT as equally open for them as it was for boys and men. This requires us to stop thinking of the gender equality norm as productive for our everyday life and rather starting to ask how our everyday activities can contribute to enacting gender equality. Meanwhile, many women are *doing IT for themselves*, some with support from Pippi, as if the strongest girl in the world and her disregard for reserved spaces and unfamiliar skills can help to shatter the masculine norm of IT.

REFERENCES

Abbate, J. (2012). *Recoding Gender. Women's Changing Participation in Computing*. MIT Press.

Ahmed, S. (2012). *On Being Included: Racism and Diversity in Institutional Life*. Duke University Press.

Ahmed, S. (2016). *Living a Feminist Life*. Duke University Press.

Aivaloglou, E., & Hermans, F. (2019). Early Programming Education and Career Orientation: The Effects of Gender, Self-efficacy, Motivation and Stereotypes. In *SIGCSE '19: Proceedings of the 50th ACM Technical Symposium on Computer Science Education* (pp. 679–685). Association for Computing Machinery (ACM). https://doi.org/10.1145/3287324.3287358

Alshahrani, A., Ross, I., & Wood, M. I. (2018). Using Social Cognitive Career Theory to Understand Why Students Choose to Study Computer Science. *Proceedings of the 2018 ACM Conference on International Computing Education Research*, 205–214. https://doi.org/10.1145/3230977.3230994

Andresen, S. (2020). Being Inclusive When Talking about Diversity: How Teachers Manage Boundaries of Norwegianness in the Classroom. *Nordic Journal of Comparative and International Education (NJCIE), 4*(3–4), 26–38.

Ardies, J., Dierickx, E., & Van Strydonck, C. (2021). My Daughter a STEM-Career? 'Rather Not'or 'No Problem'? A Case Study. *European Journal of STEM Education, 6*(1), 14.

Armoni, M., & Gal-Ezer, J. (2014). Early Computing Education—Why? What? When? Who? *ACM Inroads, 5*(4), 54–59.

Arnold, G., Dee, H., Herman, C., Moore, S., Palmer, A., & Shah, S. (Eds.). (2021). *Women in Tech: A Practical Guide to Increasing Gender Diversity and Inclusion*. BCS Learning & Development Ltd.

© The Author(s) 2024
H. G. Corneliussen, *Reconstructions of Gender and Information Technology*, https://doi.org/10.1007/978-981-99-5187-1

Bandura, A. (1977). Self-efficacy: Toward a Unifying Theory of Behavioral Change. *Psychological Review, 84*(2), 191–215.

Barbieri, D., Caisl, J., Karu, M., Lanfredi, G., Mollard, B., Peciukonis, V., La Hoz, M. B. P., Reingardé, J., & Salanauskaité, L. (2020). *Gender Equality Index 2020: Digitalisation and the Future of Work.* European Institute for Gender Equality. https://data.europa.eu/doi/10.2839/79077

Barbieri, D., Caisl, J., Lanfredi, G., Linkeviciute, J., Mollard, B., Ochmann, J., Peciukonis, V., Reingarde, J., Kullman, M., & Thil, L. (2022). *Artificial Intelligence, Platform Work and Gender Equality.* https://eige.europa.eu/publications/artificial-intelligence-platform-work-and-gender-equality

Barbieri, D., Lelleri, R., Maxwell, K., Mollard, B., Karu, M., Salanauskaité, L., & Reingardé, J. (2018). *Women and Men in ICT: A Chance for Better Work–Life Balance—Research Note.* European Institute for Gender Equality, Publications Office of the European Union. https://doi.org/10.2839/310959

Barker, L. J., & Aspray, W. (2006). The State of Research on Girls and IT. In J. M. Cohoon & W. Aspray (Eds.), *Women and Information Technology. Research on Underrepresentation* (pp. 3–54). MIT Press.

BCS. (2019). *BCS Insights Report.* https://www.bcs.org/media/2938/insights-report-2019.pdf

Becker, B. A. (2021). What Does Saying that 'Programming is Hard' Really Say, and About Whom? *Communications of the ACM, 64*(8), 27–29.

Berger, P., & Luckmann, T. ([1966] 1991). *The Social Construction of Reality. A Treatise in the Sociology of Knowledge.* Penguin Books.

Bleijenbergh, I. (2018). Transformational Change towards Gender Equality: An Autobiographical Reflection on Resistance During Participatory Action Research. *Organization, 25*(1), 131–138. https://doi.org/10.1177/1350508417726547

Blum, L., Frieze, C., Hazzan, O., & Dias, M. B. (2007). A Cultural Perspective on Gender Diversity in Computing. In C. J. Burger, E. G. Creamer, & P. S. Meszaros (Eds.), *Reconfiguring the Firewall. Recruiting Women to Information Technology across Cultures and Continents* (pp. 109–133). A K Peters, Ltd.

Borgonovi, F., Centurelli, R., Dernis, H., Grundke, R., Horvát, P., Jamet, S., Keese, M., Liebender, A.-S., Marcolin, L., Rosenfeld, D., & Squicciarini, M. (2018). *Bridging the Digital Gender Divide: Include, Upskill, Innovate.* OECD Directorate for Science, Technology and Innovation (STI). https://www.oecd.org/digital/bridging-the-digital-gender-divide.pdf

Branch, E. H. (Ed.). (2016). *Pathways, Potholes, and the Persistence of Women in Science: Reconsidering the Pipeline.* Lexington Books.

Bray, F. (2007). Gender and Technology. *Annual Review of Anthropology, 36*, 37–53.

Bryant, A. (2021). Continual Permutations of Misunderstanding: The Curious Incidents of the Grounded Theory Method. *Qualitative Inquiry, 27*(3–4), 397–411. https://doi.org/10.1177/1077800420920663

Budgeon, S. (2015). Individualized Femininity and Feminist Politics of Choice. *European Journal of Women's Studies, 22*(3), 303–318.

Buland, T. H., Mordal, S., & Mathiesen, I. H. (2020). *Utdannings-og yrkesrådgiving og sosialpedagogisk rådgiving i norsk skole anno 2020: En kartlegging av ressurser og arbeidsoppgaver til rådgiving i skolen i Norge.* NTNU Samfunnsforskning.

Bury, R. (2011). She's Geeky: The Performance of Identity among Women Working in IT. *International Journal of Gender, Science and Technology, 3*(1). http://genderandset.open.ac.uk/index.php/genderandset/article/view/113

Buse, K. (2018). Women's Under-representation in Engineering and Computing: Fresh Perspectives on a Complex Problem. *Frontiers in Psychology, 9.* https://doi.org/10.3389/978-2-88945-493-8

Castañeda-Navarrete, J., Corneliussen, H. G., & Pisciella, A. (2023). *Gender, Digital Transformation, and Artificial Intelligence: A Background Paper for the United Nations Industrial Development Organization.* UNIDO Report June 2023. United Nations Industrial Development Organization. https://hub. unido.org/sites/default/files/publications/GENDER%2C%20DIGITAL%20 TRANSFORMATION%20AND%20AI%20REPORT.pdf

Charles, M., & Bradley, K. (2006). A Matter of Degrees: Female Underrepresentation in Computer Science Programs Cross-Nationally. In J. M. Cohoon & W. Aspray (Eds.), *Women and Information Technology. Research on Underrepresentation* (pp. 183–203). MIT Press.

Charles, M., & Thébaud, S. (2018). *Gender and STEM: Understanding Segregation in Science, Technology, Engineering and Mathematics (special issue of Social Sciences).* MDPI.

Charmaz, K. (2006). *Constructing Grounded Theory: A Practical Guide Through Qualitative Research.* Sage Publications Ltd.

Charmaz, K. (2017). The Power of Constructivist Grounded Theory for Critical Inquiry. *Qualitative Inquiry, 23*(1), 34–45. https://doi.org/10.1177/1077800416657105

Chavatzia, T. (2017). *Cracking the Code: Girls' and Women's Education in Science, Technology, Engineering and Mathematics (STEM).* UNESCO.

Cheryan, S., Lombard, E. J., Hudson, L., Louis, K., Plaut, V. C., & Murphy, M. C. (2019). Double Isolation: Identity Expression Threat Predicts Greater Gender Disparities in Computer Science. *Self and Identity, 19*(4), 412–434. https://doi.org/10.1080/15298868.2019.1609576

Cheryan, S., Master, A., & Meltzoff, A. N. (2015). Cultural Stereotypes as Gatekeepers: Increasing Girls' Interest in Computer Science and Engineering by Diversifying Stereotypes. *Frontiers in Psychology, 6*(49), 1–8.

Cheryan, S., Plaut, V. C., Davies, P. G., & Steele, C. M. (2009). Ambient Belonging: How Stereotypical Cues Impact Gender Participation in Computer Science. *Journal of Personality and Social Psychology, 97*(6), 1045–1060. https://doi.org/10.1037/a0016239

Cheryan, S., Plaut, V. C., Handron, C., & Hudson, L. (2013). The Stereotypical Computer Scientist: Gendered Media Representations as a Barrier to Inclusion for Women. *Sex Roles, 69*(1–2), 58–71.

Cheryan, S., Ziegler, S. A., Montoya, A. K., & Jiang, L. (2017). Why are Some STEM Fields More Gender Balanced than Others? *Psychological Bulletin, 143*(1), 1.

Chow, T., & Charles, M. (2019). An Inegalitarian Paradox: On the Uneven Gendering of Computing Work around the World. In C. Frieze & J. L. Quesenberry (Eds.), *Cracking the Digital Ceiling: Women in Computing around the World* (pp. 25–45). Cambridge University Press.

Cockburn, C. (1992). The Circuit of Technology: Gender, Identity and Power. In R. Silverstone & E. Hirsch (Eds.), *Consuming Technologies: Media and Information in Domestic Spaces* (pp. 32–47). Routledge.

Cockburn, C., & Ormrod, S. (1993). *Gender and Technology in the Making.* SAGE Publications.

Cohoon, J. M., & Aspray, W. (Eds.). (2006). *Women and Information Technology: Research on Underrepresentation.* MIT Press.

Corneliussen, H. G. (2011). *Gender–Technology Relations: Exploring Stability and Change.* Palgrave Macmillan. https://doi.org/10.1057/9780230354623

Corneliussen, H. G. (2021a). Unpacking the Nordic Gender Equality Paradox in ICT Research and Innovation. *Feminist Encounters: A Journal of Critical Studies in Culture and Politics, 5*(2), Article 25. https://doi.org/10.20897/femenc/11162

Corneliussen, H. G. (2021b). Women Empowering Themselves to Fit into ICT. In E. Lechman (Ed.), *Technology and Women's Empowerment* (pp. 46–62). Routledge.

Corneliussen, H. G., & Prøitz, L. (2016). Kids Code in a Rural Village in Norway: Could Code Clubs Be a New Arena for Increasing Girls' Digital Interest and Competence? *Information, Communication & Society, 19*(1), 95–110. https://doi.org/10.1080/1369118X.2015.1093529

Corneliussen, H. G., & Seddighi, G. (2019). "Må vi egentlig ha flere kvinner i IKT?" Diskursive forhandlinger om likestilling i IKT-arbeid. *Tidsskrift for kjønnsforskning, 43*(4), 273–287. https://doi.org/10.18261/issn.1891-1781-2019-04-03

Corneliussen, H. G., & Seddighi, G. (2020a). Employers' Mixed Signals to Women in IT: Uncovering How Gender Equality Ideals are Challenged by Organizational Context. In P. Kommers & G. C. Peng (Eds.), *Proceedings for the International Conference ICT, Society, and Human Beings 2020* (pp. 41–48). IADIS Press.

Corneliussen, H. G., & Seddighi, G. (2020b). The Challenge of Implementing the National Gender Equality Norm in IT Organizations. *IADIS International Journal on Computer Science and Information Systems, 15*(2), 1–14.

Corneliussen, H. G., & Seddighi, G. (2022). Unconventional Routes into ICT Work: Learning from Women's Own Solutions for Working around Gendered Barriers. In G. Griffin (Ed.), *Gender Inequalities in Tech-Driven Research and Innovation: Living the Contradiction* (pp. 56–75). Bristol University Press.

Corneliussen, H. G., Seddighi, G., & Dralega, C. A. (2019). Women's Experience of Role Models in IT: Landmark Women, Substitutes, and Supporters. In Ø. Helgesen, E. Nesset, G. Mustafa, P. Rice, & R. Glavee-Geo (Eds.), *Modeller: Fjordantologien 2019* (pp. 375–395). Universitetsforlaget. https://doi.org/10.18261/9788215034393-2019-18

Corneliussen, H. G., Seddighi, G., Iqbal, A., & Andersen, R. (2023). Artificial Intelligence in the Public Sector in Norway: A Hop-On- Hop-Off Journey. In *Symposium on AI, Data and Digitalization (SAIDD 2023)* (pp. 17–21). Western Norway Research Institute.

Corneliussen, H. G., Seddighi, G., Simonsen, M., & Urbaniak-Brekke, A. M. (2021). *Evaluering av Jenter og teknologi*. VF-rapport 3/2021.

Corneliussen, H. G., Seddighi, G., Urbaniak-Brekke, A. M., & Simonsen, M. (2021). Factors Motivating Women to Study Technology: A Quantitative Survey among Young Women in Norway. In P. Kommers & M. Macedo (Eds.), *Proceedings of the IADIS International Conferences ICT, Society and Human Beings; Web Based Communities and Social Media 2021; and e-Health 2021* (pp. 202–206). IADIS Press.

Corral-Granados, A., Rapp, A. C., & Smeplass, E. (2022). Barriers to Equality and Cultural Responsiveness in Three Urban Norwegian Primary Schools: A Critical Lens for School Staff Perceptions. *The Urban Review, 55*, 94–132. https://doi.org/10.1007/s11256-022-00642-5

Czopp, A. M., Kay, A. C., & Cheryan, S. (2015). Positive Stereotypes are Pervasive and Powerful. *Perspectives on Psychological Science, 10*(4), 451–463.

Dasgupta, N. (2011). Ingroup Experts and Peers as Social Vaccines Who Inoculate the Self-concept: The Stereotype Inoculation Model. *Psychological Inquiry, 22*(4), 231–246.

Dee, H. (2021). Computing in Schools. In G. Arnold, H. Dee, C. Herman, S. Moore, A. Palmer, & S. Shah (Eds.), *Women in Tech: A Practical Guide to Increasing Gender Diversity and Inclusion* (pp. 24–47). BCS Learning & Development Ltd.

Denning, P. (2004). The Profession of IT: The Field of Programmers Myth. *Communications of the ACM, 47*(7), 15–20.

Denning, P. J., & McGettrick, A. (2005). Recentering Computer Science. *Communications of the ACM, 48*(11), 15–19. https://doi.org/10.1145/1096000.1096018

Devillard, S., Sancier-Sultan, S., Zelicourt, A. d., & Kossoff, C. (2016). Women Matter 2016: Reinventing the Workplace to Unlock the Potential of Gender Diversity. McKinsey & Co. https://shorturl.at/pBEFG

Dick, P. (2004). Resistance to Diversity Initiatives. In R. Thomas, A. Mills, & J. H. Mills (Eds.), *Identity Politics at Work: Resisting Gender and Gendering Resistance* (pp. 67–84). Routledge.

DiSalvo, B., Guzdial, M., & Bruckman, A. (2014). Saving Face While Geeking Out: Video Game Testing as a Justification for Learning Computer Science. *The Journal of the Learning Sciences, 23*(3), 272–315. https://doi.org/ 10.1080/10508406.2014.893434

Dixon-Fyle, S., Dolan, K., Hunt, V., & Prince, S. (2020). *Diversity Wins: How Inclusion Matters*. McKinsey & Company.

Dralega, C. A., & Corneliussen, H. G. (2018). Manifestations and Contestations of Hegemony in Video Gaming by Immigrant Youth in Norway. In A. A. Williams, Tsuria, R., Robinson, L. and Khilnani, A. (Eds.), *Media and Power in International Contexts: Perspectives on Agency and Identity (Studies in Media and Communications, Vol. 16).* (pp. 153–169). Emerald Publishing Limited. https://doi.org/10.1108/S2050-20602018000016011

Eccles, J. (2009). Who Am I and What Am I Going to Do with My Life? Personal and Collective Identities as Motivators of Action. *Educational Psychologist, 44*(2), 78–89. https://doi.org/10.1080/00461520902832368

Eccles, J. (2011). Gendered Educational and Occupational Choices: Applying the Eccles et al. Model of Achievement-Related Choices. *International Journal of Behavioral Development, 35*(3), 195–201.

Eccles, J. S. (2015). Gendered Socialization of STEM Interests in the Family. *Journal of Gender, Science and Technology, 7*(2), 116–132. http://genderand-set.open.ac.uk/index.php/genderandset/article/view/419/692

Eccles, J. S., & Wigfield, A. (2002). Motivational Beliefs, Values, and Goals. *Annual Review of Psychology, 53*(1), 109–132.

Ekeland, A., Pajarinen, M., & Rouvinen, P. (2015). *Computerization and the Future of Jobs in Norway*. Statistics Norway.

Ellingsæter, A. L. (2014). Kjønnsessensialisme—segregeringens evighetsmaskin? In L. Reisel & M. Teigen (Eds.), *Kjønnsdeling og etniske skiller på arbeids-markedet* (pp. 86–106). Gyldendal Akademisk.

Engström, S. (2018). Differences and Similarities between Female Students and Male Students that Succeed within Higher Technical Education: Profiles Emerge Through the Use of Cluster Analysis. *International Journal of Technology and Design Education, 28*(1), 239–261. https://doi.org/10.1007/ s10798-016-9374-z

Ensmenger, N. L. (2012). *The Computer Boys Take Over: Computers, Programmers, and the Politics of Technical Expertise*. MIT Press.

European Commission. (2013). Women active in the ICT sector – Final report. Directorate-General for the Information Society and Media. https://data. europa.eu/doi/10.2759/27822

European Commission. (2021a). *2021 Report on Gender Equality in the EU*. Publications Office of the European Union.

European Commission. (2021b). *2030 Digital Compass: The European Way for the Digital Decade*.

European Commission. (2021c). *She Figures 2021: Gender in Research and Innovation*. European Union.

European Union. (2021). *ICT Specialists: The Skills Gap Hinders Growth in the EU Countries*. Digital Skills & Jobs Platform, EU.

Eurostat. (2021a). *ICT Specialists by Sex 2010–2020*. Retrieved May 10, 2022, from https://ec.europa.eu/eurostat/statistics-explained/index.php?title=ICT_specialists_in_employment#ICT_specialists_by_sex

Eurostat. (2021b). *ICT Specialists in Employment: Statistics Explained*.

Eurostat. (2021c). *Index of the Number of Persons Employed as ICT Specialists and Total Employment, EU, 2011–2020*.

Faulkner, W. (2009). Doing Gender in Engineering Workplace Cultures. II. Gender In/authenticity and the In/visibility Paradox. *Engineering Studies, 1*(3), 169–189. https://doi.org/10.1080/19378620903225059

Faulkner, W. (2014). Can Women Engineers be "Real Engineers" and "Real Women"? In W. Ernst & I. Horwath (Eds.), *Gender in Science and Technology: Interdisciplinary Approaches* (pp. 187–203). transcript Verlag.

Fenstermaker, S., & West, C. (2002). "Doing Difference" Revisited: Problems, Prospects, and the Dailouge in Feminist Theory. In S. Fenstermaker & C. West (Eds.), *Doing Gender, Doing Difference* (pp. 205–216). Routledge.

Foss, E. S. (2020). SSB Analyse: kvinner og realfag: Gode skoleresultater—liten endring i yrkesvalg. (2020/02). https://www.ssb.no/utdanning/artikler-og-publikasjoner/gode-skoleresultater-liten-endring-i-yrkesvalg

Frieze, C., & Quesenberry, J. (2015). *Kicking Butt in Computer Science: Women in Computing at Carnegie Mellon University*. Dog Ear Publishing.

Frieze, C., & Quesenberry, J. L. (2019). *Cracking the Digital Ceiling: Women in Computing Around the World*. Cambridge University Press.

Gerson, S. A., Morey, R. D., & van Schaik, J. E. (2022). Coding in the Cot? Factors Influencing 0–17s' Experiences with Technology and Coding in the United Kingdom. *Computers & Education, 178*, 104400. https://doi.org/10.1016/j.compedu.2021.104400

Glaser, B. G., & Strauss, A. L. ([1999] 2017). *The Discovery of Grounded Theory: Strategies for Qualitative Research*. Routledge. https://doi.org/10.4324/9780203793206

González-Pérez, S., Mateos de Cabo, R., & Sáinz, M. (2020). Girls in STEM: Is It a Female Role-Model Thing?. *Frontiers in Psychology, 11*, 2204. https://doi.org/10.3389/fpsyg.2020.02204

Griffin, G. (Ed.). (2022). *Gender Inequalities in Tech-Driven Research and Innovation: Living the Contradiction*. Bristol University Press.

Griffin, G., & Vehviläinen, M. (2021). The Persistence of Gender Struggles in Nordic Research and Innovation. *Feminist Encounters: A Journal of Critical Studies in Culture and Politics*, 5(2), 1–28. https://doi.org/10.20897/femenc/11165

Griffiths, M., & Moore, K. (2010). "Disappearing Women": A Study of Women Who Left the UK ICT Sector. *Journal of Technology Management & Innovation*, 5(1), 95–107.

Grover, S., Pea, R., & Cooper, S. (2014). Remedying Misperceptions of Computer Science among Middle School Students. In *Proceedings of the 45th ACM Technical Symposium on Computer Science Education* (pp. 343–348). ACM. https://doi.org/10.1145/2538862.2538934

Guzdial, M. (2015). Learner-Centered Design of Computing Education: Research on Computing for Everyone. *Synthesis Lectures on Human-Centered Informatics*, 8(6), 1–165.

Hacker, S. (1989). *Pleasure, Power, and Technology. Some Tales of Gender, Engineering, and the Cooperative Workplace*. Unwin Hyman.

Haraway, D. (1991). A Cyborg Manifesto. In D. Haraway (Ed.), *Simians, Cyborgs, and Women. The Reinvention of Nature* (pp. 149–181). Free Association Books.

Hayes, C. C. (2010). Gender Codes: Prospects for Change. In T. J. Misa (Ed.), *Gender Codes: Why Women are Leaving Computing* (pp. 265–273). IEEE Computer Society and John Wiley & Sons, Inc.

Hernes, H. M. (1987). *Welfare State and Woman Power. Essays in State Feminism*. Norwegian University Press.

Holst, C., Skjeie, H., & Teigen, M. (2019). Likestillingspolitikk og europeisk integrasjon. In C. Holst, H. Skjeie, & M. Teigen (Eds.), *Europeisering av nordisk likestillingspolitikk* (pp. 11–34). Gyldendal Akademisk.

Holter, Ø. G., & Snickare, L. (Eds.). (2021). *Likestilling i akademia–fra kunnskap til endring*. Cappelen Damm Akademisk/NOASP (Nordic Open Access Scholarly Publishing).

Holtzblatt, K., & Marsden, N. (2022). *Retaining Women in Tech: Shifting the Paradigm*. Morgan & Claypool Publishers.

Hunt, V., Prince, S., Dixon-Fyle, S., & Yee, L. (2018). Delivering Through Diversity. *McKinsey & Company Report*, 3, 2018.

Hyde, J. S. (2005). The Gender Similarities Hypothesis. *American Psychologist*, 60(6), 581.

Insight Intelligence. (2022). *Unga Kvinnor och IT 2022*.

Jacobs, J., Ahmad, S., & Sax, L. (2017). Planning a Career in Engineering: Parental Effects on Sons and Daughters. *Social Sciences*, 6(1), 2.

Jethwani, M. M., Memon, N., Seo, W., & Richer, A. (2016). "I Can Actually Be a Super Sleuth": Promising Practices for Engaging Adolescent Girls in Cybersecurity Education. *Journal of Educational Computing Research*, 55(1), 3–25. https://doi.org/10.1177/0735633116651971

Jones, L. K., & Hite, R. L. (2020). Expectancy Value Theory as an Interpretive Lens to Describe Factors That Influence Computer Science Enrollments and Careers for Korean High School Students. *Electronic Journal for Research in Science & Mathematics Education, 24*(2), 86–118.

Kaiser, C. R., Major, B., Jurcevic, I., Dover, T. L., Brady, L. M., & Shapiro, J. R. (2013). Presumed Fair: Ironic Effects of Organizational Diversity Structures. *Journal of Personality and Social Psychology, 104*(3), 504.

Kanter, R. M. ([1977] 1993). *Men and Women of the Corporation.* Basic Books.

Kenny, E. J., & Donnelly, R. (2020). Navigating the Gender Structure in Information Technology: How Does This Affect the Experiences and Behaviours of Women? *Human Relations, 73*(3), 326–350.

Kleif, T., & Faulkner, W. (2003). "I'm No Athlete [but] I Can Make This Thing Dance!"—Men's Pleasures in Technology. *Science, Technology, & Human Values, 28*(2, Spring), 296–325.

Laclau, E., & Mouffe, C. (1985). *Hegemony and Socialist Strategy: Towards a Radical Democratic Politics.* Verso.

Lagesen, V. A., Pettersen, I., & Berg, L. (2021). Inclusion of Women to ICT Engineering—Lessons Learned. *European Journal of Engineering Education, 47*, 1–16.

Landström, C. (2007). Queering Feminist Technology Studies. *Feminist Theory, 8*(1), 7–26.

Lang, C., Fisher, J., Craig, A., & Forgasz, H. (2020). Computing, Girls and Education: What We Need to Know to Change How Girls Think about Information Technology. *Australasian Journal of Information Systems, 24.* https://doi.org/10.3127/ajis.v24i0.1783

Larsen, E., Moss, S. M., & Skjelsbæk, I. (2021). *Gender Equality and Nation Branding in the Nordic Region.* Taylor & Francis.

Lewis, C. M., Anderson, R. E., & Yasuhara, K. (2016). "I Don't Code All Day": Fitting in Computer Science When the Stereotypes Don't Fit. *Proceedings of the 2016 ACM Conference on International Computing Education Research,* Melbourne, VIC, Australia. https://doi.org/10.1145/2960310.2960332

Lien, M. I. (2021). Mellom universalisme og feminisme: en sosiologisk analyse av kjønn og omsorgsbegrepet i sykepleiefaget. *Tidsskrift for samfunnsforskning, 62*(3), 233–250. https://doi.org/10.18261/issn.1504-291X-2021-03-02x

Lyon, L. A., & Green, E. (2020). Women in Coding Boot Camps: An Alternative Pathway to Computing Jobs. *Computer Science Education, 30*(1), 102–123.

Margolis, J., & Fisher, A. (2002). *Unlocking the Clubhouse. Women in Computing.* MIT Press.

Mariscal, J., Mayne, G., Aneja, U., & Sorgner, A. (2019). Bridging the Gender Digital Gap. *Economics, 13*(1). https://doi.org/10.5018/economics-ejournal.ja.2019-9

Martinsson, L., & Griffin, G. (2016). *Challenging the Myth of Gender Equality in Sweden*. Policy Press.

Massey, D. (1996). *Space, Place and Gender*. Polity Press.

Master, A., Cheryan, S., & Meltzoff, A. N. (2014). Reducing Adolescent Girls' Concerns about STEM Stereotypes: When Do Female Teachers Matter? *Revue internationale de psychologie sociale, 27*(3), 79–102.

Master, A., Cheryan, S., & Meltzoff, A. N. (2016). Computing Whether She Belongs: Stereotypes Undermine Girls' Interest and Sense of Belonging in Computer Science. *Journal of Educational Psychology, 108*(3), 424–437.

Master, A., Cheryan, S., Moscatelli, A., & Meltzoff, A. N. (2017). Programming Experience Promotes Higher STEM Motivation among First-grade Girls. *Journal of Experimental Child Psychology, 160*, 92–106.

Master, A., & Meltzoff, A. N. (2020). Cultural Stereotypes and Sense of Belonging Contribute to Gender Gaps in STEM. *International of Gender, Science and Technology, 12*(1), 152–198.

McKinney, V. R., Wilson, D. D., Brooks, N., O'Leary-Kelly, A., & Hardgrave, B. (2008). Women and Men in the IT Profession. *Communications of the ACM, 51*(2), 81–84.

McKinsey & Company and Pivotal Ventures. (2018). *Rebooting Representation: Using CSR and Philanthropy to Close the Gender Gap in Tech*. Tech Report 2018. Retrieved September 5, 2023, from https://www.rebootrepresentation.org/wp-content/uploads/Rebooting-Representation-Report.pdf

Michell, D., Szorenyi, A., Falkner, K., & Szabo, C. (2017). Broadening Participation Not Border Protection: How Universities Can Support Women in Computer Science. *Journal of Higher Education Policy and Management, 39*(4), 406–422.

Minelgaite, I., Sund, B., & Stankeviciene, J. (2020). Understanding the Nordic Gender Diversity Paradox. *TalTech Journal of European Studies, 10*(1), 40–57.

Misa, T. J. (2010a). Gender Codes: Lessons from History. In T. J. Misa (Ed.), *Gender Codes: Why Women are Leaving Computing* (pp. 251–263). IEEE Computer Society and John Wiley & Sons, Inc.

Misa, T. J. (Ed.). (2010b). *Gender Codes: Why Women are Leaving Computing*. IEEE Computer Society and John Wiley & Sons, Inc.

Mordal, S., Buland, T., & Mathiesen, I. H. (2020). Career Guidance in Norwegian Primary Education: Developing the Power of Dreams and the Power of Judgement. In E. Hagaseth Haug, T. Hooley, J. Kettunen, & R. Thomsen (Eds.), *Career and Career Guidance in the Nordic Countries* (Vol. 9). Brill.

Moss-Racusin, C. A., Sanzari, C., Caluori, N., & Rabasco, H. (2018). Gender Bias Produces Gender Gaps in STEM Engagement. *Sex Roles, 79*(11), 651–670. https://doi.org/10.1007/s11199-018-0902-z

Nentwich, J. C., & Kelan, E. K. (2014). Towards a Topology of "Doing Gender": An Analysis of Empirical Research and Its Challenges. *Gender, Work and Organization, 21*(2), 121–134. https://doi.org/10.1111/gwao.12025

NIFU. (2021). *Realistiske forventninger? Sluttrapport fra evalueringen av Tett på realfag. Nasjonal strategi for realfag i barnehagen og grunnopplæringen (2015–2019).* Nordisk institutt for studier av innovasjon, forskning og utdanning NIFU. https://hdl.handle.net/11250/2836261

NOU. (2019: 19). *Jenterom, gutterom og mulighetsrom—Likestillingsutfordringer blant barn og unge.* Kulturdepartementet.

OECD. (2016). *PISA 2015 Results (Volume I). Excellence and Equity in Education.*

OECD. (2021). Norway. In *Education at a Glance 2021: OECD Indicators.* OECD Publishing.

Oldenziel, R. (1999). *Making Technology Masculine: Men, Women and Modern Machines in America, 1870–1945.* Amsterdam University Press.

Olsen, B. M. O., & Wendt, K. (2023). *Økt andel kvinner blant professorene i 2022.* Statistics Norway. Retrieved September 5, 2023, from https://www.ssb.no/teknologi-og-innovasjon/forskning-og-innovasjon-i-naeringslivet/statistikk/forskerpersonale/artikler/okt-andel-kvinner-blant-professorene-i-2022

Palmer, A. (2021). Attracting Women into Tech. In G. Arnold, H. Dee, C. Herman, S. Moore, A. Palmer, & S. Shah (Eds.), *Women in Tech: A Practical Guide to Increasing Gender Diversity and Inclusion* (pp. 133–174). BCS Learning & Development Ltd.

Pantic, K., & Clarke-Midura, J. (2019). Factors that Influence Retention of Women in the Computer Science Major: A Systematic Literature Review. *Journal of Women and Minorities in Science and Engineering, 25*(2), 119–145. https://doi.org/10.1615/JWomenMinorScienEng.2019024384

Pawluczuk, A., Lee, J., & Gamundani, A. M. (2021). Bridging the Gender Digital Divide: An Analysis of Existing Guidance for Gender Digital Inclusion Programmes' Evaluations. *Digital Policy, Regulation and Governance, 23*(3), 287–299. https://doi.org/10.1108/DPRG-11-2020-0158

Petray, T., Doyle, T., Howard, E., Morgan, R., & Harrison, R. (2019). Re-Engineering the "Leaky Pipeline" Metaphor: Diversifying the Pool by Teaching STEM "by Stealth". *International Journal of Gender, Science and Technology, 11*(1), 20. http://genderandset.open.ac.uk/index.php/genderandset/article/view/582/1027

Prottsman, K. (2014). Computer Science for the Elementary Classroom. *ACM Inroads, 5*(4), 60–63.

Puwar, N. (2004). *Space Invaders: Race, Gender and Bodies Out of Place.* Berg.

Puwar, N. (n.d.). Space Invaders: Insiders'/Outsiders' Ontological Complicity. *Manipulations.*

Quirós, C. T., Morales, E. G., Pastor, R. R., Carmona, A. F., Ibáñez, M. S., & Herrera, U. M. (2018). *Women in the Digital Age*. Luxembourg: Publications Office of the European Union, Issue.

Reisel, L., Skorge, Ø. S., & Uvaag, S. (2019). *Kjønnsdelte utdannings- og yrkesvalg. En kunnskapsoppsummering*. Institutt for samfunnsforskning.

Rohatgi, A., Scherer, R., & Hatlevik, O. E. (2016). The Role of ICT Self-efficacy for Students' ICT Use and Their Achievement in a Computer and Information Literacy Test. *Computers & Education, 102*, 103–116. https://doi.org/10.1016/j.compedu.2016.08.001

Sáinz, M., & Eccles, J. S. (2012). Self-concept of Computer and Math Ability: Gender Implications across Time and Within ICT Studies. *Journal of Vocational Behavior, 80*(2), 486–499.

Sammet, K., & Kekelis, L. (2016). *Changing the Game for Girls in STEM: Findings on High Impact Programs and System-Building Strategies*. Techbridge.

Schiro, E. C. (2022). *Norsk mediebarometer 2021*. S. sentralbyrå.

Seibel, S., & Veilleux, N. (2019). Factors Influencing Women Entering the Software Development Field Through Coding Bootcamps vs. Computer Science Bachelor's Degrees. *Journal of Computing Sciences in Colleges, 34*(6), 84–96.

Sevin, R., & Decamp, W. (2016). From Playing to Programming: The Effect of Video Game Play on Confidence with Computers and an Interest in Computer Science. *Sociological Research Online, 21*(3), 14–23. https://doi.org/10.5153/sro.4082

Sey, A., & Hafkin, N. (2019). *Taking Stock: Data and Evidence on Gender Equality in Digital Access, Skills and Leadership*. United Nations University, Tokyo.

Sharma, K., Torrado, J. C., Gómez, J., & Jaccheri, L. (2021). Improving Girls' Perception of Computer Science as a Viable Career Option Through Game Playing and Design: Lessons from a Systematic Literature Review. *Entertainment Computing, 36*, 100387. https://doi.org/10.1016/j.entcom.2020.100387

Simonsen, M., & Corneliussen, H. G. (2020). What Can Statistics Tell About the Gender Gap in ICT? Tracing Men and Women's Participation in the ICT Sector Through Numbers. In D. Kreps, T. Komukai, T. V. Gopal, & K. Ishii (Eds.), *Human-Centric Computing in a Data-Driven Society* (pp. 379–397). Springer International Publishing.

Smith, S., Taylor-Smith, E., Fabian, K., Barr, M., Berg, T., Cutting, D., Paterson, J., Young, T., & Zarb, M. (2020). Computing Degree Apprenticeships: An Opportunity to Address Gender Imbalance in the IT Sector? *2020 IEEE Frontiers in Education Conference (FIE)*, Uppsala, Sweden, 2020, pp. 1–8. https://doi.org/10.1109/FIE44824.2020.9274144.

Snickare, L., & Holter, Ø. G. (2021). Likestilt ubalanse? In Ø. G. Holter & L. Snickare (Eds.), *Likestilling i akademia—fra kunnskap til endring* (pp. 19–48). Cappelen Damm Akademisk. https://doi.org/10.23865/noasp.143

Søndergaard, D. M. (2002). Poststructuralist Approaches to Empirical Analysis. *Qualitative Studies in Education, 15*(2), 187–204.

Sørensen, K. H. (2011). Changing Perspectives on Gender and Technology: From Exclusion to Inclusion. In K. H. Sørensen, W. Faulkner, & E. Rommes (Eds.), *Technologies of Inclusion. Gender in the Information Society* (pp. 41–61). Tapir Academic Press.

Spieler, B., Grandl, M., Ebner, M., & Slany, W. (2019). Computer Science for All: Concepts to Engage Teenagers and Non-CS Students in Technology. In *European Conference on Games Based Learning*, Reading, UK.

Statistics Norway. (2018). *Women and Men in Norway*. Statistics Norway.

Statistics Norway. (2022). *Norwegian Media Barometer*.

Steele, C. M., & Aronson, J. (1997). A Threat in the Air: How Stereotypes Shape Intellectual Identity and Performance. *American Psychologist, 52*(6), 613–629.

Steine, F. S., Gunnes, H., & Wendt, K. (2020). *Gender Balance in Research—December 2020: Gender Balance among Researchers in Norwegian Academia*.

Stoet, G., & Geary, D. C. (2018). The Gender–Equality Paradox in Science, Technology, Engineering, and Mathematics Education. *Psychological Science, 29*(4), 581–593. https://doi.org/10.1177/0956797617741719

Stout, J. G., Dasgupta, N., Hunsinger, M., & McManus, M. A. (2011). STEMing the Tide: Using Ingroup Experts to Inoculate Women's Self-concept in Science, Technology, Engineering, and Mathematics (STEM). *Journal of Personality and Social Psychology, 100*(2), 255.

Strauss, A., & Corbin, J. (2008). *Basics of Qualitative Research: Techniques and Procedures for Developing Grounded Theory*. Sage Publications. https://doi.org/10.4135/9781452230153

Sultan, U. N., Axell, C., & Hallström, J. (2019). Girls' Engagement with Technology Education: A Scoping Review of the Literature. *Design and Technology Education: An International Journal, 24*(2), 20–41.

Sund, B. (2015). Just an Illusion of Equality? The Gender Diversity Paradox in Norway. *Beta, 29*(2), 157–183.

Tænketanken DEA. (2019). *Hvordan får vi STEM på lystavlen hos børn og unge?—Og hvilken rolle spiller køn for interesseskabelsen?* Retrieved May 10, 2021, from https://www.datocms-assets.com/22590/1589284030-pixi-stempaaly tavlenhosboernogunge.pdf

Talks, I., Edvinsson, I., & Birchall, J. (2019). *Programmed Out: The Gender Gap in Technology in Scandinavia*. Plan International Norway.

Tandrayen-Ragoobur, V., & Gokulsing, D. (2021). Gender Gap in STEM Education and Career Choices: What Matters? *Journal of Applied Research in Higher Education, ahead-of-print*(ahead-of-print). https://doi.org/10.1108/JARHE-09-2019-0235

Teigen, M., & Skjeie, H. (2017). The Nordic Gender Equality Model. In O. Knutsen (Ed.), *The Nordic Models in Political Science. Challenged, but Still Viable?* (pp. 125–148). Fagbokforlaget.

The European Institute for Gender Equality. (2017). *Economic Benefits of Gender Equality in the EU. How Gender Equality in STEM Education Leads to Economic Growth*. European Institute for Gender Equality (EIGE), European Commission.

The Norwegian Directorate for Education and Training. (2023). *Fagvalg i videregående skole*.

The Norwegian Universities and Colleges Admission Service. (2022). *Applicant and Admission Statistics*.

Thomas, T., & Allen, A. (2006). Gender Differences in Students' Perceptions of Information Technology as a Career. *Journal of Information Technology Education: Research, 5*(1), 165–178.

Traphagen, K., & Traill, S. (2014). *How Cross-sector Collaborations are Advancing STEM Learning*. Noyce Foundation.

Trauth, E., & Connolly, R. (2021). Investigating the Nature of Change in Factors Affecting Gender Equity in the IT Sector: A Longitudinal Study of Women in Ireland. *MIS Quarterly, 45*(4), 2055–2100.

Trauth, E. A., & Quesenberry, J. L. (2007). Gender and the Information Technology Workforce: Issues of Theory and Practice. In P. Yoong & S. Huff (Eds.), *Managing IT Professionals in the Internet Age* (pp. 18–36). Idea Group Publishing. https://doi.org/10.4018/978-1-59140-917-5.ch002

Turkle, S. (1988). Computational Reticence. Why Women Fear the Intimate Machine. In C. Kramarae (Ed.), *Technology and Women's Voices. Keeping in Touch* (pp. 41–61). Routledge & Kegan Paul.

United Nations Industrial Development Organization. (2021). *Industrial Development Report 2022: The Future of Industrialization in a Post-Pandemic World*.

Vainionpää, F., Kinnula, M., Iivari, N., & Molin-Juustila, T. (2019). Gendering and Segregation in Girls' Perceptions of IT as a Career Choice: A Nexus Analytic Inquiry. In A. Siarheyeva, C. Barry, M. Lang, H. Linger, & C. Schneider (Eds.), *Information Systems Development: Information Systems Beyond 2020 (ISD 2019 Proceedings)* (pp. 1–12). ISEN Yncréa Méditerranée.

Vekiri, I. (2013). Information Science Instruction and Changes in Girls' and Boy's Expectancy and Value Beliefs: In Search of Gender-Equitable Pedagogical Practices. *Computers & Education, 64*, 104–115.

Vitores, A., & Gil-Juárez, A. (2016). The Trouble with 'Women in Computing': A Critical Examination of the Deployment of Research on the Gender Gap in Computer Science. *Journal of Gender Studies, 25*(6), 666–680.

Vrieler, T., Nylén, A., & Cajander, Å. (2020). Computer Science Club for Girls and Boys—A Survey Study on Gender Differences. *Computer Science Education*, 1–31. https://doi.org/10.1080/08993408.2020.1832412

Vrieler, T., & Salminen-Karlsson, M. (2022). A Sociocultural Perspective on Computer Science Capital and its Pedagogical Implications in Computer Science Education. *ACM Transactions on Computing Education (TOCE)*, *22*(4), 1–23.

Wajcman, J. (1991). *Feminism Confronts Technology*. Polity Press.

Wajcman, J. (2004). *TechnoFeminism*. Polity Press.

Wajcman, J. (2010). Feminist Theories of Technology. *Cambridge Journal of Economics, 34*(1), 143–152.

Watts, J. H. (2009). "Allowed into a Man's World": Meanings of Work–Life Balance: Perspectives of Women Civil Engineers as 'Minority' Workers in Construction. *Gender, Work & Organization, 16*(1), 37–57.

West, C., & Zimmerman, D. H. (1987). Doing Gender. *Gender & Society, 1*(2), 125–151. https://doi.org/10.1177/0891243287001002002

West, C., & Zimmerman, D. H. (2009). Accounting for Doing Gender. *Gender & Society, 23*(1), 112–122.

Wong, B., & Kemp, P. E. (2018). Technical Boys and Creative Girls: The Career Aspirations of Digitally Skilled Youths. *Cambridge Journal of Education, 48*(3), 301–316.

World Economic Forum. (2020a). *The Future of Jobs Report 2020*. https://www3.weforum.org/docs/WEF_Future_of_Jobs_2020.pdf

World Economic Forum. (2020b). *The Global Gender Gap Report 2020*. World Economic Forum.

Wynn, A., & Correll, S. (2017). Gendered Perceptions of Cultural and Skill Alignment in Technology Companies. *Social Sciences, 6*(2), 45. http://www.mdpi.com/2076-0760/6/2/45

Yates, J., & Plagnol, A. C. (2022). Female Computer Science Students: A Qualitative Exploration of Women's Experiences Studying Computer Science at University in the UK. *Education and Information Technologies, 27*(3), 3079–3105. https://doi.org/10.1007/s10639-021-10743-5